The Essential Engineer

The Essential Engineer

WHY SCIENCE ALONE WILL NOT SOLVE OUR GLOBAL PROBLEMS

Henry Petroski

Alfred A. Knopf · New York

2010

THIS IS A BORZOI BOOK
PUBLISHED BY ALFRED A. KNOPF

A portion of this work originally appeared
in *Civil Engineering Magazine*.

Library of Congress Cataloguing-in-Publication Data
Petroski, Henry.
The essential engineer : why science alone will not
solve our global problems / by Henry Petroski.—1st ed.
p.   cm.
"A Borzoi book."
Includes bibliographical references and index.
ISBN 978-0-307-27245-4
1. Technological innovations—Popular works.   2. Engineering—Popular works.
3. Technology and civilization—Popular works.   4. Research—Popular works.
5. Problem solving—Popular works.   6. Science—Popular works.   I. Title.
T173.8.P467 2010
620-dc22   2009021216

Manufactured in the United States of America
First Edition

To Ash Green

Contents

Preface   ix

1. Ubiquitous Risk   3
2. Engineering Is Rocket Science   18
3. Doctors and Dilberts   29
4. Which Comes First?   45
5. Einstein the Inventor   59
6. Speed Bumps   71
7. Research and Development   93
8. Development and Research   105
9. Alternative Energies   124
10. Complex Systems   156
11. Two Cultures   173
12. Uncertain Science and Engineering   184
13. Great Achievements and Grand Challenges   198
14. Prizing Engineering   218

Notes   231
Illustration Credits   259
Index   261

Preface

I am a faithful reader of "Science Times," the Tuesday feature section of *The New York Times*, but I have often wished that it were called something more precisely descriptive, for the section is about more than just science. On a typical Tuesday about half of the section's content relates to health; sometimes there are even stories about engineering. Both medicine and engineering do use scientific knowledge and methods to solve relevant problems, but neither of them is simply an applied science. In fact, the practices of medicine and engineering are more like each other than either is like unqualified science: medical doctors and engineers both welcome all the relevant science they can muster, but neither can wait for complete scientific understanding before acting to save a life or create a new life-saving machine.

Nevertheless, the word *science* is commonly understood to include medicine, engineering, and high technology. "Science" is clearly a useful shorthand for a wide range of activities, but it also obscures the differences between them. It gives science a primacy that it may or may not deserve. This book seeks to illuminate the differences between science and engineering and thereby clarify their respective roles in the worlds of thought and action, of knowing and doing. In particular, it focuses on how they can interact to define and solve some of the most interesting and pressing problems of our time: how to address climate change, clean and renewable energy, and other global threats and challenges.

· · ·

Some of the material in this book appeared first, in somewhat different form, in my regular columns in *American Scientist*, the magazine of Sigma Xi, the Scientific Research Society; and in *ASEE Prism*, the magazine of the American Society for Engineering Education. I first drafted the manuscript in Maine during the summer of 2008, drawing many of my examples and case studies from contemporary and near-contemporary events, as reported in *The New York Times* and elsewhere. Among the resources that were most helpful to me in identifying relevant stories were the daily summaries of news distributed via e-mail by Sigma Xi ("Science in the News"), the American Society for Engineering Education ("First Bell"), and the American Society of Civil Engineers ("ASCE SmartBrief"). I am grateful to the anonymous scanners of the news who produce these excellent digests. The final version of the manuscript was completed in Durham, North Carolina, during the fall of 2008 and winter of 2009, at which time I added further examples and made the material current as of then. However, I believe that the underlying principles upon which this book is based are independent of when it was written. It is a book about science and engineering, those uniquely human endeavors that will continue to shape and not follow the news of the day.

As always, I am indebted to my wife, Catherine Petroski, for taking time from her own work to be my first reader and thoughtful assistant. And I am pleased that my relationship with the publishing house of Alfred A. Knopf continues even after the retirement of my longtime editor and friend, Ashbel Green, to whom this book is dedicated. This is my tenth book with Knopf, and my first with my new editor, Andrew Miller. It has been a pleasure getting to know Andrew and working with him on this project. And, as I have come to expect, my experience with Ellen Feldman and the entire production team at Knopf has been superb.

The Essential Engineer

1

Ubiquitous Risk

Our lives and those of our children and grandchildren are constantly at risk. Hardly a day passes, it seems, when there is not a story on television or in the newspaper about some new threat to our health and safety. If it is not toys decorated with lead-based paint, then it is drugs—not just pharmaceuticals but something as commonplace as toothpaste—containing adulterated ingredients, or even milk contaminated with industrial chemicals that found its way into candy sold around the globe.[1]

Risk and reassurance are two key considerations of the activities of science, engineering, invention, and technology—collectively often referred to simply as "science" or "science and technology." Whatever they are called, they play a critical role in modern civilization, being essential for the advancement of society and the protection of our quality of life. It is these human disciplines associated with discovery and design that help separate the good from the dangerous on the farm and in the factory, at home and at the office, and on battlefields and frontiers. While science and technology can be misused and become the source of ruin, we would be at even greater risk from tainted products and contagious diseases were it not for the benevolent use of what are among the achievements that make us most distinctly human. If science and technology are two-edged swords, they are also the essential weapons in detecting and managing everyday risk.

The bad milk that caused so much consternation a couple of years ago originated in China, which is among the largest exporters of food and food ingredients in the world. In order to increase quantities and thus realize greater profit, unscrupulous participants in the food supply chain misused chemical engineering to water down and adulterate milk. However, diluted milk, being lower in protein, can easily be detected by standardized testing employing well-established technology. But by adding inexpensive melamine, a chemical rich in nitrogen that is used in producing fertilizer and plastics, the adulterated milk could be made to register a higher protein level. Some of the tainted milk found its way into baby formula, causing tens of thousands of children to become ill, with at least six infants dying. This happened because melamine does not dissolve easily in the body and in higher concentrations can produce kidney stones and lead to kidney failure. The widespread presence of melamine in Chinese food products, including cookies and yogurt, led to worldwide recalls. Melamine had also been used as a cheap filler in pet food, causing many cats and dogs to become seriously ill. The chemical was additionally suspected to have been used in other animal feed, which caused chickens to produce melamine-tainted eggs. China promised to crack down on such practices—going so far as to sentence to death some of those responsible for the criminal activity—but the incident prompted a nagging skepticism that soon there could be some other tainted import that we would have to worry about.[2]

The Chinese milk scandal is a striking example of the use and misuse of science and technology and of the tragic consequences that can result. In themselves, science and technology are neutral tools that help us understand the world and allow us to work with its resources. People, however, are not necessarily neutral participants, and they can use their scientific understanding and technical prowess for good or ill. It may be that those who added melamine to diluted milk thought they were only being clever exploiters of chemistry. The unfortunate consequences of their actions were, of course, beyond mere venality, and ironically, the very same science and technology that served as tools of deception were also used to uncover the plot. Like risk itself, science and technology and their effects are ubiquitous.[3]

It is not just potentially harmful products from abroad that can give us pause. Not long ago E. coli–contaminated spinach from California proved to be the culprit in the deaths of three people and the illnesses of hundreds of Americans who trusted domestically grown and harvested produce. A few years later, salmonella-tainted tomatoes were believed initially to be responsible for causing hundreds of people in dozens of states to become ill. For a while, the root of the problem, which spread through forty-one states and affected more than a thousand people, was believed to be in Florida, or maybe Mexico. When no source was found in either of those agricultural locations, however, the public was told that perhaps tomatoes were not the source after all. Maybe it was fresh jalapeños—or something else. Six weeks after advising people not to eat tomatoes, the U.S. Food and Drug Administration lifted the advisory without reaching any definite conclusion about the origin of the salmonella. It was not that science and technology were inadequate to the task. It was that there were no reliable data trails pointing to the various hands through which the bad food had passed on its way to the supermarket. When the guilty bacterium was finally found in a Texas distribution plant, its ultimate origin could not be traced. Unfortunately, such elastic and inconclusive warnings inure us to risk.[4]

Not long after the tomato/jalapeño incident, peanut products containing salmonella were traced to a processing plant in Georgia. In the years preceding the discovery, the plant had been cited repeatedly by the state department of agriculture for health violations, ranging from unclean food preparation surfaces to dirty and mildewed walls and ceilings. On numerous occasions, when the company's own testing detected salmonella in its products, they were retested with negative results and the products were shipped. It was only after a salmonella outbreak was traced to peanut butter from the plant that it was shut down by the Food and Drug Administration and two years' worth of peanut butter products were recalled—after the company was given an opportunity to approve the wording of the recall statement. A selective interpretation of scientific test results and a casual enforcement of technical regulations can imperil millions of people. Such incidents threaten the reputation that science and technology once held for objectivity and are likely to bring increased calls for tightened regulation.[5]

In the wake of the salmonella scares, the Food and Drug Administration approved the use of radiation on fresh vegetables like lettuce and spinach to rid them of bacteria. An editorial in *The New York*

Times praised the move, noting that astronauts have long eaten irradiated meat, and that other treated foods, like poultry and shellfish, had produced no detectable adverse effects on those consumers who had tried them. Of course, there remain a great number of people who cringe at the idea of eating anything that has been exposed to radiation, and it is likely going to be a long time before the practice can be expected to become the norm. Nevertheless, it is such technological advances, which ultimately owe their existence to science and engineering research and development, that can bring an overall reduction in risks of all kinds, including those involved in activities as common and essential as eating. [6]

In modern times, systems of commercial competitiveness and government regulation have provided a good measure of checks and balances against undue risk, but the failings of human nature can interfere with the proper functioning of those protective social structures. Science and engineering can be called upon to develop new means of defining safe limits of contaminants and toxins and can devise new instruments and methods for detecting unsafe products, but the ultimate reduction in risk from everyday things is more a matter of vigilance and enforcement than of technology. It is imperative that positive results for salmonella and other contaminants be taken seriously and treated responsibly by the private food industry. If there continues to be life-threatening disregard for consumer health and safety, it is likely that increased government oversight will be imposed.

Sometimes new technology—even that encouraged by law—brings with it new risks, and we are forced to confront the unthought-of consequences of a seemingly good idea. In recent years, the increased use of crops like corn in the manufacture of biofuels intended to ease our dependence on foreign oil pinched the food supply and caused prices to rise. To avoid this problem, nonfood crops have increasingly been proposed for making second-generation green fuels. But biologists have warned that certain reeds and wild grasses known to botanists as "invasive species" and to gardeners as "weeds" would have a high likelihood of overtaking nearby fields, presenting serious threats to the ecology and economy of a region. Investors in the fast-growing worldwide biofuels industry naturally reject such doomsday scenarios, but the risk is a real one. The European Union has been especially bullish on biofuels, with plans to use them for 10 percent of the fuel needed for transportation by 2020. However, it has become increasingly clear that agricultural efforts undertaken to help meet that goal were leading to deforestation in remote regions, thereby contributing to climate change and affecting food prices worldwide. In fact, taking into account production and transportation costs, biofuels may do more harm to the global environment than fossil fuels. New technologies can certainly harbor even newer surprises.[7]

Another potentially risky new technology is the much-touted nanotechnology, which concerns itself with substances and structures whose size is on the scale of atoms and molecules. Nanotubes, already put to use in something so familiar as a tennis racket, are essentially ultra-tiny rolled-up sheets of carbon that are employed in the production of materials much stronger and lighter than steel. Unfortunately, the tubes are shaped like microscopic needles, a property that has caused scientists to speculate that they might present the same health hazard as asbestos, whose fibers have a similar shape. Since nanotubes date only from the early 1990s, their risk as possible carcinogens is not yet fully known.[8]

Nanomaterials of all kinds are increasingly being used in a wide variety of consumer products. Nanoparticles of silver, which are known to be very effective in killing bacteria, have been incorporated into clothing fabrics as a means of preventing the buildup of bacteria that produce undesirable odors in such articles as socks. This obvious advantage may prove to come at a price, however, for as the clothing is worn, washed, and disposed of, the nanosilver is leached out and released to the environment, where it can accumulate and do uncertain harm. For example, washed-out silver particles might destroy bacteria that are an integral part of the filtering process in municipal wastewater systems. A British royal commission on environmental pollution warned that "the potential benefits of nanomaterials meant that the rise in their use had far outstripped the knowledge of the risks they might pose."[9]

We are not home free even in hospitals. Epidemiologists have estimated that one in twenty-two patients contracts a hospital infection. And, according to an Institute of Medicine report published in 2000, medical error in American hospitals has been blamed for 44,000 to 98,000 unnecessary deaths per year, making inadvertent deaths due to "preventable hospital error" the number eight cause of death annually—above fatalities due to motor vehicle accidents, breast cancer, and AIDS. We must risk our life in trying to save it.[10]

Yet there is little outrage. It is not necessary that we accept an inordinate level of risk as an inevitable by-product of technology. For example, the odds of being killed on an airliner are as long as one in ten million; it is not uncommon for an entire year to pass without a single fatal commercial airplane crash in the United States. This outstanding record has been accomplished by taking seriously rules and regulations, procedures and processes—sometimes to the inconvenience and anger of impatient passengers. If a plane has a mechanical problem, it does not take off until the problem has been diagnosed and resolved. There was considerable disruption to air traffic when American Airlines was forced to cancel hundreds of flights following a disagreement with the Federal Aviation Administration over the safety inspection of essential cables bundled together in wheel wells. Many resigned fliers sighed, "Better safe than sorry."[11]

When a commercial airliner does crash and passengers are killed, it is instant news, in part because it is as rare an occurrence as a man-bites-dog story. But how often do we hear on the national news of a hospital patient dying because of an improperly administered drug or an infection contracted in the course of routine medical treatment—not to mention a misdiagnosis or an overdose of improperly prescribed pills? Unless the patient is a celebrity or a well-known politician, such incidents remain the private tragedies of family and close friends. If the medical and its ancillary professions had in place—and assiduously followed—rules, regulations, and procedures as stringent as those of the aircraft industry, it is likely that the rate of deaths due to medical errors would be greatly reduced.

It seems that the more common the occurrence of something, the more we tend to accept it as part of the unavoidable risk of living. Even perceived risk can all but immobilize the perceiver. While there are certainly people who fear going into the hospital lest they never leave it alive, only the unusual individual will not seek supervised medical treatment when it is needed. Anecdotally, at least, there also seem to be many people who avoid flying because of their fear of never leaving the plane alive. But when a travel emergency arises, most will relent and go to the airport. Risk numbers may support a fear of hospitals, but they simply do not support a fear of flying. Irrational fears can be nonetheless compelling.

In fact, the more remote the chance of something happening, the more we also seem to fear it. It is as if the sheer unfamiliarity of the thing—perhaps because its unfamiliarity has been magnified by our very avoidance of it, or perhaps because it is something theretofore not experienced by mankind—notches up the perceived risk to emphasize the need for precaution. A global catastrophe was feared when the first astronauts to land on the Moon would return to Earth. What if they brought back with them some deadly lunar microorganism that could cause the entire population of the planet to fall fatally ill? The risk was considered "extremely remote" but real enough to quarantine the returning astronauts until they were deemed not to be contagious. Another global catastrophe was feared when the first hydrogen bomb was tested, some scientists expressing genuine concern that there was an extremely small but real possibility that the explosion would ignite the atmosphere and destroy all life on the planet. In both cases, the risk could have been eliminated entirely by not going ahead with the new technology, but compelling geopolitical motives prevailed.[12]

More recently, some physicists expressed concern about the Large Hadron Collider, the enormous international particle accelerator built and operated by CERN, the European Organization for Nuclear Research. Located near Geneva, and straddling the border between Switzerland and France, the collider has been described variously as "the biggest machine ever built," "the most powerful atom-smasher," and "the largest scientific experiment in history." The purpose of the mostly underground machine is to send protons, which contain collections of elementary particles known as hadrons, into various targets in the hope of observing never-before-seen subatomic particles or uncovering never-before-conceived aspects of the universe. The fear was that using the collider, which is designed to operate at unprecedented energy levels, could "spawn a black hole that could swallow Earth" or trigger some other cataclysmic event. An unsuccessful lawsuit to block the start-up of the device alleged that there was "a significant risk" involved and that the "operation of the Collider may have unintended consequences which could ultimately result in the destruction of our planet." The director general of CERN countered with a news release declaring that the machine was safe and any suggestion of risk was "pure fiction." Scientists involved with the project were determined to go ahead with it, even though some of them received death threats. In the end, the collider forces prevailed, and test operations began at low energy levels in the late summer of 2008.[13]

It is necessary to begin slowly with large systems like the collider, for such an enormously complicated machine brings with it a multitude of opportunities for predictable surprises with unpredictable consequences. Just thirty-six hours after it was started up, during which time many beams of protons were successfully sent through the tubes of the collider, it had to be shut down because it was believed that one of its electrical transformers failed. It was replaced and test operations began again, but something was still not right. It turned out that there was a leak in the liquid helium cooling system; the cause was attributed to a single poorly soldered connection—just one of ten thousand made during construction—between two of the machine's fifty-seven magnets, which produced a hot spot that led to the breach. The collider operates in a supercooled state—near absolute zero temperature—which meant that it had to be warmed up before the leak could be repaired. The entire process of warming up, repairing the leak, and then cooling down to operating temperature again was at first expected to take at least a couple of months. That proved to be an optimistic estimate, for most of the magnets were damaged by a buildup of pressure associated with the helium leak and had to be replaced. The machine was in fact shut down for about a year. Such are the risks associated with complex technology, but scientists and engineers expect such setbacks and tend to take them in stride. In time, after all of the bugs had been ironed out, the collider was expected to operate as designed—and with no fatal consequences to scientists, engineers, or planet Earth anticipated.[14]

But rare cataclysmic events do occur, and they have been described as "the most extreme examples of risk." It is believed that about four billion years ago an object the size of Mars struck the Earth and disgorged material that became the Moon. Scientists have also hypothesized that billions of years ago a meteor the size of Pluto—packing energy equivalent to as many as 150 trillion megatons of TNT and producing the solar system's largest crater—struck Mars and thereby caused that planet's unbalanced shape. This asymmetry was discovered in the 1970s by observations made from Viking orbiters, which detected that there was a two-mile difference in altitude between the red planet's upper third and bottom two-thirds. Other scientists favor the hypothesis that internal forces are responsible for the lopsidedness.[15]

In 1994, our solar system was the scene of an unusually spectacular event, one described as "recorded history's biggest show of cosmic violence." Parts of Comet Shoemaker-Levy 9 rained down on Jupiter, producing enormous (Earth-sized) fireballs that "outshone the planet" and were easily visible through a small telescope. It has been estimated that the energy released in the collision exceeded that of all the nuclear weapons in the world. The question soon arose, What if a meteor, comet, or asteroid were to be on a collision course with our own planet? And what if we saw it coming? Could anything be done about it?[16]

No matter how low the probability, it was obvious that the consequences of such an event could be devastating. Legitimate concerns led to a congressionally mandated study by the National Aeronautics and Space Administration, which was to assess the dangers of such a collision and how it might be anticipated and avoided. NASA outlined a proposal involving a worldwide network of telescopes through which the skies could be watched to provide an early warning of anything on a collision course with Earth. The focus should clearly be on those so-called near-Earth objects that have the potential for doing the most harm. According to NASA, a worst-case scenario would occur if a large comet or asteroid hit with the energy of thousands of nuclear warheads exploding at the same time in the same location. This would enshroud the planet in dust, blocking out sunlight and disrupting the climate to such an extent that it would be the end of civilization as we know it. Such a collision is believed to have occurred 65 million years ago and led to the extinction of the dinosaurs. However, should such an event be anticipated far enough in advance, it might be possible for scientists and engineers to track the object and devise an interception plan, whereby Earth could be saved.[17]

Still, skeptics abounded. There were those who thought that giving serious attention to a "wildly remote" possibility was "laughably paranoid." Others doubted that a monitoring plan could gain sufficient political support to get into the federal budget. But with the NASA report fresh in its mind when the Shoemaker-Levy comet encountered Jupiter, the House Science Committee voted to charge the space agency with identifying and cataloguing "the orbital characteristics of all comets and asteroids greater than one kilometer in diameter in orbit around the sun that cross the orbit of the Earth."[18]

Even a relatively small asteroid could do significant damage if it came close enough to Earth before exploding in the atmosphere, however. This is what is believed to have happened in Siberia a century ago. The so-called Tunguska Incident is known to have occurred through eyewitness accounts and physical evidence. A person living about forty miles south of the location of the occurrence recalled seeing on June 30, 1908, how "the sky split in two and fire appeared high and wide over the forest." His body became unbearably hot on the side facing north. Then there was the sound of a strong thump, and he was thrown backward. The Earth shook, wind blew, and windows were shattered. People from as far as hundreds of miles away reported hearing the blast. Seismometers recorded the equivalent of a Richter 5 earthquake. An estimated eighty million trees were knocked down over an area of eight hundred square miles. In the 1920s, the Russian mineralogist Leonid Alekseyevich Kulik led research expeditions to the extensive site and discovered that the felled trees radiated from a central spot, but he found no fragments of a meteorite in the vicinity.[19]

The scientific consensus appears to be that an asteroid measuring

Tens of millions of trees were flattened throughout hundreds of square miles of forest in 1908, when an asteroid exploded in Earth's atmosphere above the Stony Tunguska River, which is located in Siberia.

maybe 150 feet across—though some scientists think it could have been much smaller—exploded about five miles above the surface of the Earth. The explosion carried the force of about fifteen megatons of TNT, which would make it a thousand times more powerful than the atomic bomb dropped on Hiroshima. Some scientists and politicians used the one-hundredth anniversary of the Tunguska event to call attention to their belief that not enough was being done to defend Earth against asteroids and comets. NASA now maintains its Near Earth Object Program office at the Jet Propulsion Laboratory to identify potential threats, but, according to one scientist, "the greatest danger does not come from the objects we know about but from the ones we haven't identified."[20]

One evening in the early fall of 2008, someone watching the sky from an observatory near Tucson, Arizona, noticed an incoming object. By the next morning, three other skywatchers—located in

California, Massachusetts, and Italy—had confirmed that an asteroid provisionally designated 2008 TC3 (indicating the year of its discovery and a coded reference to when in that year the discovery took place) was speeding toward our planet. The collective information about its trajectory enabled astronomers to compute and thus predict that the following day the object would collide with Earth's atmosphere in the sky above a tiny Sudan village. The impact occurred at the predicted place and within minutes of the predicted time, which NASA had publicized about seven hours before the actual collision. According to the program manager, this represented "the first time we were able to discover and predict an impact before the event," even though such an impact takes place about once every three months—making it far from a rare occurrence. The atmospheric impact energy of 2008 TC3 was estimated to be the equivalent of one to two thousand tons of TNT, from which it could be inferred that the asteroid had a diameter of about ten feet. Fortunately, this rock from space disintegrated when it hit Earth's atmosphere and any fragments that may have fallen to the ground did no harm.[21]

As of mid-2009, the NASA center had on its list more than 6,000 objects that might one day strike Earth, with about 750 of them being large enough to do considerable damage. The critical diameter appears to be under a mile, but known incoming objects just one-sixth of critical size would prompt a warning from NASA to evacuate the endangered area. Such a warning could be issued several days beforehand. In the case of the Sudan event, however, asteroid 2008 TC3 was simply "too small and dark to be discovered until it was practically upon Earth," and so it was not on NASA's watch list. Those on the list, especially the larger objects, will allow for plenty of warning—and possibly even the opportunity for engineers to do something about them. And there are protective measures that can be taken.[22]

In the early stages of the asteroid-tracking effort, which is referred to as Spaceguard, the science-fiction author, inventor, and futurist Arthur C. Clarke supported the concept. In commentary in *The New York Times*, he described how Earth-threatening bodies might be deflected from their target. Referring to some of his own novels, in which such tasks were undertaken, Clarke outlined three possible approaches. The first he termed the "brute force approach: nuke the beast." The equivalent of a billion tons of high explosives could split the incoming rock into fragments, some of which might still do damage to places on Earth, but not to the cataclysmic extent that the whole body would have.[23]

Clarke's second means was to send up astronauts to mount thruster rockets on the asteroid. Even only a slight nudge from these thrusters, exerted over a sufficiently long period of time, would change the object's trajectory just enough so that the cumulative effect would be to miss the Earth entirely. Since it takes the Earth about six minutes to move a distance equal to its own diameter, slowing down or speeding up a threatening asteroid's arrival time by just six minutes could make the difference between whether it strikes or misses Earth.[24]

Finally, Clarke described an "even more elegant solution" involving the mounting of a metal foil mirror on the foreign body, thereby employing the tiny but persistent pressure of sunlight to push the body into a deflected orbit. Because of the small forces involved, however, such a scheme would need years or even decades of continuous action to make a difference, but, given enough lead time, it could work. Clarke was engaging not in the observational and predictive thinking of scientists but in the conceptual and constructive thinking of engineers.[25] This is a key point, and it is a topic to which we shall return.

Not long after the incident of the Jupiter fireworks, a Harvard astronomer announced that an asteroid was on a collision course with Earth, and an alarmingly close encounter would occur thirty years hence. He alerted the world that the recently discovered asteroid, designated 1997 XF11, should be expected to come within thirty thousand miles of our planet at about 1:30 p.m. on October 26, 2028. Because thirty thousand miles is less than four times the diameter of Earth, and given the uncertainty of such a long-range prediction, there appeared to be a reasonable chance that there might actually be a collision. However, within a week of the announcement, another astronomer—at the Jet Propulsion Laboratory—who had used additional data relating to the asteroid to recalculate its orbit, found that it would "come no closer than 600,000 miles and had no chance of hitting the planet."[26]

It is not uncommon for different scientists looking at the same phenomenon to reach different conclusions; this is what makes it difficult for laypersons to sort out the truth and risk relating to everything from medical procedures to global climate change. In the immediate wake of the contradictory predictions of the asteroid's encounter with Earth, a group of fifteen astronomers formed an expert committee that could estimate what risk to Earth would be posed by a threatening asteroid.

The following year, at a meeting of the International Astronomical Union, astronomers adopted the Torino Impact Hazard Scale, which takes into account the energy involved as well as the probability that a particular asteroid will strike Earth. The Torino scale, ranging from 0 for objects that will miss Earth to 10 for those capable of causing global destruction, thus takes into account both risk magnitude and consequence, thereby enabling a more meaningful comparison of distinct events. The asteroid that killed off the dinosaurs would have rated a 10, but no recorded object would have earned more than a 1. The Tunguska event, known largely through anecdotal and circumstantial evidence observed after the fact, is not considered to have been "recorded" in the astronomical sense.[27]

The New York Times, editorializing about the new scale, observed that asteroid 1997 XF11, whose predicted encounter with Earth caused so much embarrassment to the astronomical community, would have dropped considerably in its rating over the few days it took to recalculate its orbit. Of course, computational errors can be off in both directions, meaning that there is risk even in our reliance on quantifications of risk.[28]

Just as predicting landfall for a hurricane moving over the Gulf of Mexico requires constant updating as more information and data become available, so pinpointing where a body hurtling through space will strike Earth necessarily changes with time. Scientists can give it their best shot to predict risk within a certain margin of error, but the ultimate answer to the question of where something will strike can be known with certainty only at the last moment. It thus involves a judgment call to decide when to stop hoping that the scientific tracking and predicting will tell us we are safe, and when to begin taking steps to alter the course of nature. This is where science hands the problem over to engineering. Science is about knowing; engineering about doing. Or, as I once heard in a lecture on climate change, "scientists warn, engineers fix." But it is not always easy to distinguish science from engineering or scientists from engineers, for there can be considerable overlap in their aims and methods. This book strives to clarify the often hazy distinction between science and engineering, between scientists and engineers, thereby making clearer what they can and cannot do about ameliorating global risks that have been termed "planetary emergencies"—such as global climate change—that appear to threaten us and our world. Understanding the distinctions better enables more informed judgments and decisions relating to public policy issues such as those concerning management of risk and the allocation of resources for research and development.[29]

2

Engineering Is Rocket Science

The lead story on the front page of *The New York Times* bore the headline "Nuclear Weapons Engineer Indicted in Removal of Data." However, on the inside page on which the article was continued, a related story about the arrest of Wen Ho Lee, the former Los Alamos National Laboratory staff member who had come under suspicion for spying, bore the headline "Lee's Defenders Say the Scientist Is a Victim of a Witch Hunt Against China." As interested as I was in the content of these stories, I became fixated on the two descriptions of Lee in the headlines. Did the *Times* really consider *engineer* and *scientist* to be synonymous?[1]

The business of creating headlines for a newspaper of record is not, or at least should not be, taken lightly. Lewis Jordan, a former editor of the *Times*, once prepared a primer for the school classroom on how the news is written and edited. In the section on headline writing, he stated:

> The serious newspaper insists that headlines, like stories, be clear, grammatical, impartial and accurate. But the headline must also be interesting. . . . One way to be interesting in a headline is to be specific rather than general. It is better to say "Swimming Champion Saved from Drowning" than to say "Noted Athlete Saved from Drowning."[2]

By this criterion, one can infer that the *Times* headline writers— if they did not make an error—did indeed believe, at least at the time the two headlines about Wen Ho Lee were written, that a "nuclear weapons engineer" was a special kind of scientist. That belief apparently changed over the course of the next two weeks, however, for in follow-up stories on the declared innocence and defense of Lee by himself and his friends and colleagues, he was consistently described in the *Times* as a "nuclear weapons scientist," which may indeed have been his job title, even though all his degrees were earned in mechanical engineering. Despite the overlap and confusion in terminology, engineering and science remain distinct human endeavors.[3]

Science and technology reporting in newspapers and magazines is nothing new, nor is the confusion between engineers and scientists, or between science and engineering. In his primer, Jordan identified the "science story" as a distinct category of reporting and had this to say about it: "When Dr. Jonas E. Salk developed his polio vaccine, there were reporters equipped to write intelligently about that. When the Russians and the Americans fired their first earth satellites, there were reporters who could tell the reader why the satellites went into orbit and why they stayed there."[4] But telling the why of satellites does not explain the essential how. That the second, if not the first, was in fact an engineering story apparently escaped even the critical eye and ear of editor Jordan.

One of the examples of science stories that Jordan offered was from a September 14, 1959, edition of the *Times*, which reported that a Soviet rocket had hit the Moon to become the first man-made object to reach another cosmic body. A Moscow radio announcer, who had interrupted a program of classical music, was reported to have said that "the moon strike was an outstanding achievement of Soviet science and engineering and had opened a new page in space research." However, the long *Times* story gave scant acknowledgment of the contribution of engineering and engineers; it was dominated by the words "science" and "scientists," who were quoted to explain various aspects of the feat. Why has there been confusion between science and engineering—and frequent exclusion of the latter—in our newspapers and in common parlance?[5]

Articulating the differences between the engineer and the scientist, between engineering and science, should not take a rocket scientist, though one is often invoked when the most vexing of problems arises. Theodore von Kármán might be said to have personified the rocket scientist. Born in Budapest, Hungary, in 1881, he moved to the United States in 1930, when he joined the faculty of the California Institute of Technology and became director of that institution's Guggenheim Aeronautical Laboratory. He began to conduct rocket (or "jet") propulsion experiments on the site of what in 1944 was to be established as the Jet Propulsion Laboratory, which was taken over by the National Aeronautics and Space Administration in 1958. Among von Kármán's research achievements was an analysis of the alternating double row of vortices that occurs in a fluid stream behind a bluff body—that is, one having a blunt back side with corners that do not allow streamlines to come together gracefully. This configuration of vortices came to be known as a "Kármán vortex street." It was von Kármán's management experience and technical achievement, at least in part, that in 1940 got him appointed to the committee charged with looking into the causes of the infamous Tacoma Narrows Bridge collapse, whose downfall was related to a trailing vortex street. His contentious relationship with the bridge engineers on that committee no doubt helped form, or at least reinforce, his opinions about the nature of and the distinctions between science and engineering.[6]

Although often honored as a scientist—he received the National Medal of Science from President John F. Kennedy—von Kármán referred to himself as an engineer and had evidently given enough thought to the difference between a scientist and an engineer to be confident in articulating the oft-quoted and oft-paraphrased distinction attributed to him: *a scientist studies what is; an engineer creates what never was.* By extension, *science is the study of what is; engineering is the creation of what never was.* This is a basically reasonable distinction, but, like all tidy dichotomies, it can present difficulties when applied to the practical world. Thus it was ironic that when a commemorative postage stamp honoring von Kármán was issued in 1992, he could not be labeled simply as a "scientist" or an "engineer." Much to the chagrin of some engineers, he was identified on the stamp proper as an "aerospace scientist." But in a brief biographical note on the selvage of the sheet of stamps, he was described as an "aerodynamicist and engineer." (To confuse things still further, he was also identified as the "architect of the space age," but this is not the place to explore the distinction between engineer and architect.)[7]

In fact, even in their most basic professional activity, scientists can act like engineers (and vice versa), at least according to von Kármán's definition. Chemists regularly synthesize new compounds, and biologists create new strains of plants and animals that do not exist in nature. In other words, scientists can do engineering (as engineers can do science). This would also be the case, for example, when a scientist proposes a new hypothesis about the origin of the universe or the laws of nature. If the hypothesis is truly new, then by definition it "never was" before being articulated by the scientist. Even though the hypothesis may address among the purest of scientific concerns, such as the origin and nature of the universe, the mere fact that in its formulation it is a creation of an individual scientist may be said to put it in the category of engineering—at least according to von Kármán's distinction. Such "theory engineering" or "scientific engineering" shares with "pure engineering" the characteristic that it represents a creative act of invention.[8]

Theodore von Kármán was identified as a scientist on this 1992 commemorative U.S. postage stamp and as an engineer in the biographical note that appeared on the selvage of the sheet of stamps.

Albert Einstein, certainly thought of as a prototypical scientist, effectively held this view. He criticized the physicist and philosopher Ernst Mach for having "thought that somehow theories arise by means of *discovery* and not by means of *invention*," emphasizing himself the words of distinction. According to the historian of technology Thomas P. Hughes, to Einstein invention was the manipulation not only of things but also of concepts. He believed that "an artifact was a materialized concept" and that "a hard-and-fast line between technology and science simply did not exist."[9]

What complicates labeling engineering and science according to von Kármán's distinction is the fact that the universe of "what is" is a constantly changing collection of concepts, objects, and things. No one would argue that naturally occurring flora and fauna, geological formations, and planets and stars are not properly objects of study for science. But what of telescopes and rockets, those things that might be enlisted not only to probe to understand the universe but also to save its Earth from destruction by an asteroid? Such things, commonly known as inventions, do not exist until they are conceived and constructed, and even then there remains a certain amount of ambiguity over when they can be considered to be.

Does a rocket exist when someone first conceives of it, or when nothing more than a sketch of it has been made? The inventor of the rocket must surely be considered an engineer, or at least someone doing engineering, but when an individual works out on paper the details of making a rocket and sending it to the Moon, is that person doing science or engineering? There certainly is a lot of "rocket science" involved, in the sense that there is a great advantage to understanding the laws of physics and how they relate specifically to the thing called a rocket, but there is also considerable "aerospace engineering" involved, in that the rocket will not work unless it has the proper aerodynamic shape, fuel, propulsion system, and controls that exploit the laws of chemistry and physics. These aspects of the design cannot be taken off nature's shelf; they are pure creations of engineering.

A rocket begins in the mind's eye. Whether that eye is in the head of someone called an engineer or a scientist matters little; the act of conceiving the thing is certainly one of engineering. But rockets do not get to the Moon if they remain the private dream of an individual: to realize what never was, the concept must be communicated to others. This might be done in words, but among engineers it is more likely done, at least initially, with sketches on the back of an envelope, on a napkin, on a tablecloth, in a notebook, or on a blackboard, a whiteboard, or a computer screen. It is only when a community of engineers (perhaps including scientists sometimes acting as engineers) can see the same dream that they can begin to bring it to reality and operation through the processes known as research and development.[10]

The shape of the rocket, like the concept itself, comes from the mind of someone doing engineering. Indeed, a plethora of configurations is likely to come from the minds of the engineers who get involved in the project. Choosing among the many options involves judgment, insight, and instinct as much as engineering and science. The use of equations and formulas may be helpful only on the margins, assisting the engineers with the details. The ideal fuel, for example, is not likely to be in any chemist's catalogue, and there will be questions of power, weight, materials, corrosion resistance, fuel burn rate, and a host of other details that have to be wrestled with. Most likely, there will have to be some chemical engineering done to come up with a proper fuel, structural engineering to develop the rocket casing, mechanical engineering to design the nozzles and controls, and so forth.

As for a trajectory, in the end this demands judgment calls as to when and where on Earth the launch should take place and how the rocket should approach its target. True, the equations of space dynamics and orbital mechanics apply, but the engineer must decide on the initial and intermediate conditions and identify the constraints under which the equations must be solved. Among the key decisions in the Apollo program was exactly what sequence of spacecraft maneuvers would take place between the Earth and the Moon. There was no single right answer, as there may be to a mathematics or physics problem. The various options would have their pluses and minuses, and the trick to engineering is to play one off against another to achieve the desired end within a reasonable budget of time, cost, and risk.[11]

Building and launching the first successful rocket was no easy task, as demonstrated by the countless frustrations with design and analysis that were experienced by Robert Goddard and his team before the launch of the first liquid-fuel-propelled rocket in 1926.

The space travel theorist Hermann Oberth spent two decades working on a mathematical characterization of the various factors that affect a rocket's trajectory, finishing his work three years after Goddard's practical success. In other words, the rocket came before the mathematical solution to the problem of rocket flight. Inventors seldom have the patience of scientists.[12]

Although it recounts a naive example, the memoir *Rocket Boys*, and the film *October Sky* made from it, tells it like it was: whether achieved by teenagers or professionals, success followed many trials and failures. Once there were rockets, however, or even just concepts of rockets, they could indeed become the object of scientific investigation, as they did for Oberth. Rather than taking the stars and planets as their universe of discourse, scientists—in such a context often called "engineering scientists"—could take the created world of spacecraft as their universe of discourse. The nature of the rocket can be explored and understood in ways that might not have been fully evident to the design engineers. But this does not at all diminish the achievement of conceptualizing the first rockets, without which there would be nothing to analyze.[13]

The poet (and medical doctor) William Carlos Williams went so far as to say that there are "no ideas but in things," but our Western Platonic bias has it that ideas are superior and prerequisite to things. Hence, scientists who deal in ideas, even ideas about things, tend to be viewed as superior to engineers who deal directly in things. This point of view has no doubt contributed to the mistaken conclusion that science must precede engineering in the creative process. In fact, as the self-propelled rocket so forcefully demonstrates, the engineer can go a long way in creating what never was without a fully formed science of the thing. This was certainly the case with the pyrotechnic "arrows of flaming fire" with which the Chinese demonstrated a considerable amount of empirical achievement before the artifacts became the subject of analytic study by modern rocket scientists.[14]

Nuclear physics, the subject in which Wen Ho Lee was steeped, provides another powerful example of the distinction, and confusion, between science and engineering. The principal objective of scientists like Einstein and Niels Bohr was to understand the nature of the universe and the stuff in it. Scientists like Enrico Fermi wished to build nuclear reactors not so much to generate power as to test hypotheses about nuclear physics. It was only when the political realities of atomic power became clear and threatening that the scientists lobbied for what became the Manhattan Project, whose implicit objective was, to be sure, to test the hypothesis that a nuclear weapon could be built. But this was a distinctly different kind of hypothesis, because a nuclear weapon, which at the time did not exist as an artifact, was in itself a real and deliberate objective. The form of a bomb did not so much fall out of equations expressing the laws of physics, however, as stem from a creative act of the imagination that specified a geometry that would result in a tractable mathematical problem. The scientists were thus doing engineering, not only in the service of testing scientific hypotheses but also with the objective of producing a new artifact, something that theretofore had not existed. Collectively, they were most certainly engaged in an engineering endeavor.

It was out of the Manhattan Project that our system of national laboratories arose. Shortly after its establishment in 1946, about twenty-five miles southwest of downtown Chicago, Argonne National Laboratory was designated the principal laboratory for American nuclear reactor development—a distinctly engineering mission. Many of the laboratory's early projects and accomplishments were uniquely engineering endeavors and achievements, including reactors to generate electric power, to produce isotopes, and to conduct research. Yet, from the laboratory's founding, all of its directors have been not engineers but scientists. A history of Argonne, written on the occasion of its fiftieth anniversary, chronicled the regimes of each of the laboratory's first seven directors. The book contains over five hundred pages of text and has an exhaustive index running to more than thirty pages. There are multiple index entries for "science" and "scientists," yet there is not a single one for "engineering" or "engineers." This is especially remarkable when such a substantial part of Argonne has been dedicated to nuclear reactor engineering.[15]

Such demotions of engineering and elevations of science are not uncommon. At dinner one evening, the woman sitting to my right described a friend of the family as "a scientist and an engineer." When I asked her which he considered himself to be, she did not seem to know exactly how to respond. Her husband, sitting across the table from me, offered that their friend was educated as an engineer and so was an engineer. I then asked the woman why she considered him to be a scientist. Her response was that their friend had directed a large space telescope project, which she assumed would be the responsibility of a scientist—not an engineer. So, even though he may have been educated as an engineer, what he was doing indicated to her that he was a scientist—in a more responsible position than she expected an engineer could be.

Although there may be commonalities in principle and similarities in method, neither science nor engineering can completely subsume the other. This is not to say that self-declared or designated scientists cannot do engineering, or that engineers cannot do science. In fact, it may be precisely because they each can and do participate in each other's defining activities that scientists and engineers—and hence science and engineering—are so commonly confused. The point was graphically embodied in the von Kármán commemorative stamp.

But the anecdotal evidence of engineers' and scientists' separate and unequal status provides fodder for engineers who feel that their profession is misunderstood and undervalued. Some engineers even see the interchangeable use of "engineer" and "scientist" in newspaper headlines as more than just innocent confusion or carelessness. Could it be a manifestation of the apparent hierarchical structure between the sciences and engineering that seems to have been promoted at laboratories like Argonne? And could it be promulgated—if unwittingly—by science writers and reporters in the media, whose members have overwhelmingly studied science rather than engineering in college? For example, a careful reader of the *New York Times* coverage of Wen Ho Lee could note that it was an engineer who was indicted for removing security data but a scientist who was defended as a victim of bias.[16]

Another example, relating to embryonic stem cell research, appeared in that newspaper of record more recently. The story involved the modification of human genes, which opponents of the research claimed could lead to the development of techniques applicable to producing "designer babies" that would be more intelligent, of greater physical stature, more handsome, and better at sports—changes that were seen by critics not to be natural. The privately financed research project was defended as developing techniques of gene transfer that were important to the study of human development. The headline on the story read: "Engineering by Scientists on Embryo Stirs Criticism." Some engineers see an uncanny pattern of engineering (and engineers) being associated with bad consequences, and science (and scientists) associated with good ones.[17]

The reporting of NASA's Mars probes provides further anecdotal evidence. When in 1997 the Pathfinder mission—the earliest of the agency's "faster, cheaper, better" missions—achieved the first landing on that planet in two decades, the occasion provided an opportunity for journalists to reflect upon the nature of "a whole new generation of planetary explorers," since most of those who had worked on the pioneering interplanetary probes of the 1960s had retired before the rejuvenation of the space agency. One story, published shortly after the public had watched on television the Mars rover Sojourner leave the lander Pathfinder and explore nearby rocks and terrain, carried the headline, "A New Breed of Scientists Studying Mars Takes Control." Right beside the headline was a picture of Donna Shirley, manager of the Mars exploration office at the Jet Propulsion Laboratory, which was responsible for the mission of putting the rover on Mars, not for studying the planet. The caption beneath the photo identified her as the "designer" of the rover.[18]

Nowhere in the article was Shirley identified as the engineer that she considers herself to be. When radio contact between Sojourner and Pathfinder went dead for an extended period, there was fear that the mission would come to a premature end. The incident was not discussed until well into the story about "scientists," but when it was, it was engineers who were "not sure what caused or corrected the communications lapse." After it was fixed, the mission control team was once again jubilant, and then the news conferences were given over to the scientists. Not surprisingly, their presence increased in newspaper stories and headlines.[19]

A more recent newspaper article, one on mapping the terrain of celestial bodies, reported early efforts at characterizing the surface of planets and moons to be the work of "space engineers." Their work was described as being so primitive that the distance between landmarks was measured "with a straightedge and a crayon." Today, the article reported, with digital computers, laser instruments, and digital photography at their disposal, much better work was being done by "scientists." But it was engineers and not scientists who developed those better devices.[20]

It is not always clear what causes or corrects "communications lapses" in newspapers or social intercourse, but we should hope that they are minimized so that stereotypes and misrepresentations, whether of engineers, scientists, or any group, do not get reinforced. Unfortunately, the stereotypes, prejudices, and misconceptions have been around for some time, and so it is unlikely that they will easily disappear. A 1967 book, *The Engineer and His Profession*, reported on an anecdote that "received so much circulation during the late 1950s and early 1960s that it eventually almost assumed the status of a cliché":

Every rocket-firing that is successful is hailed as a *scientific achievement*; every one that isn't is regarded as an *engineering failure*.[21]

This kind of situation, which is still lamented by engineers today, was said to evoke the image of "an engineering profession standing petulantly aside while the scientists accept all the glory."[22] The roots of such distinctions and resentments run deep, and we shall try to uncover some of them. In the end, however, if science is to detect and engineering is to deflect an asteroid from its collision course with planet Earth, scientists and engineers will have to understand and respect each other and work together for the common good—as the best of them do, even if newspapers do not often report or acknowledge it.

3

Doctors and Dilberts

The National Academy of Engineering has reported that an opinion poll found that "the public believes engineers are not as engaged with societal and community concerns as scientists or as likely to play a role in saving lives." Furthermore, or perhaps because of this perception, a subsequent poll found that "when asked to judge the relative prestige of professions, people tend to place engineering in the middle of the pack, well below medicine, nursing, science, and teaching." It may be understandable that the public considers medicine, nursing, and teaching to be highly regarded professions, for they are represented by individuals in very direct, visible, and consequential contact with their patients and students. Few people encounter engineers on a professional person-to-person basis, and it is conceivable that a lay individual could go through life without ever dealing with an engineer in his or her capacity as an engineer. But cannot the same be said of scientists? So why are they regarded more highly than engineers? And are engineers really not "as likely to play a role in saving lives"?[1]

By the end of the nineteenth century, physicians and biological scientists had established that microorganisms ("germs") in the air and in drinking water were the cause of deadly diseases. But understanding the cause of something does not in itself prevent it from occurring. It was not until the early twentieth century, after engineers had developed effective filtration and chlorination techniques, that microorganisms were substantially removed from drinking water. The installation of filters in water supply systems alone caused the incidence of typhoid fever in the United States to drop precipitously, by about a factor of ten. The so-called sanitary engineers who were responsible for such dramatic improvements in public health have since been renamed environmental engineers, but they are no less committed to playing a significant role in saving lives by reducing pollution in the air, toxins in the soil, and contaminants in the water. Such engineers are seldom celebrated, but they are in a recent book on human waste, in which the author admitted to liking engineers: "They build things that are useful and sometimes beautiful—a brick sewer, a suspension bridge—and take little credit. They do not wear black and designer glasses like architects. They do not crow."[2]

Engineers tend to crow so little that they seldom get recognized for their work, so much of which is underground, behind architectural facades, and associated with other professionals. Many of the delivery systems for life-saving drugs and countless medical devices that are now in common use have been designed and developed by engineers working with medical doctors. These include pacemakers and regenerative medicine, which also goes by the name of tissue engineering. Scientific experiments and medical trials may expose the nature of an illness and the efficacy of a treatment strategy, but drug delivery systems, imaging machines, cardiac devices, and tools of the operating room are products of engineering and not of science or even of medicine. Unfortunately, because the nature of this kind of engineering and the achievements of the engineer versus the scientist are not widely known, even among highly educated people misconceptions abound.[3]

When scientists are thought of as superior to engineers, it seems to relate less to saving lives than to a perception of the lofty scientific pursuit versus the down-to-earth engineering achievement. Indeed, can there be any comparison between the activities of searching for new worlds and filtering dirty water? When G. Wayne Clough, a civil engineer who left his position as president of Georgia Tech to become the secretary of the Smithsonian Institution, was featured in a newspaper article, it described his career as having been in the "unglamorous field of engineering." What scientific field is described as "unglamorous"? What scientists, who tend to be more flamboyant than engineers, would be characterized as practicing in an unglamorous field? Who can recall the popular astronomer Carl

Sagan or the celebrity physicist Albert Einstein ever having been accused of being involved in an unglamorous pursuit?[4]

In its extreme form, the activity known as "pure science" is believed to be about the search for "pure truth." One scientist, who was a young researcher at the University of Cambridge in the 1930s, decades later admitted that he and his colleagues had possessed such an attitude: "We prided ourselves that the science we were doing could not, in any conceivable circumstances, have any practical use. The more firmly one could make that claim, the more superior one felt."[5]

Like art for art's sake, science for science's sake needs no external goal or referent—other than the universe itself. And like artists, who historically have depended so much on patrons for their survival, scientists, too, have come to depend on the patronage—in the form of grants and fellowships—of government, industry, and the academy for support of their research.

When the Large Hadron Collider, which has been described as a "stupendous match of science and engineering," is finally fully operational, it is expected to take research in particle physics to a new level. But some reports would have led outsiders to believe that few if any engineers were partners in developing the seventeen-mile-circumference machine, which "thousands of physicists from dozens of countries have been involved in building." Yet one more generous theoretical physicist saw the start-up of the collider, which took two decades to design and construct, as a passing of the baton: "For the engineers, the job is over; for the experimentalists, they're happy to find what they find." Indeed, the seven thousand engineers and physicists who were to be involved in the device's initial years of operation anticipated finding answers to questions about the origins of matter, space, and time. But, according to another observer, the results could be "so puzzling it could take another four decades to understand them." Still, whether the researchers find answers, more questions, or nothing, it is believed that "the adventure will have been worth it."[6]

The adventure is not always worth it for everyone, however. In the 1970s, I was a member of the staff at Argonne National Laboratory. Engineers and scientists worked side by side at the lab, often on the same ultimate project, to which by their education and training they naturally brought quite different perspectives and talents. By dint of a space crunch that was accompanying a period of funding health for the lab, I shared my first office at Argonne with a scientist. Our job titles were, respectively, "mechanical engineer" and "nuclear physicist." Although these designations may not have been of our own choosing—having been assigned to us by someone in management or in human resources—they were badges of honor for each of us in his own way. There was a certain professional pride in becoming the real engineer or scientist that we had studied to be, and we did not call ourselves the one when we knew we were the other, even though to a casual observer our activities might have seemed at times to be indistinguishable. For most of the day, we sat at identical gray steel desks and pored over reams of printouts from the same mainframe computer.

However, although our desks were in the same room, we reported to different technical managers, albeit within the same administrative division, which dealt with reactor analysis and safety. Roughly speaking, the first part of the name was considered a scientific endeavor, the second an engineering responsibility, though as always with science and engineering there was also plenty of overlap. My scientist office mate was a quiet fellow, who worked silently with large computer programs referred to as codes. These were designed to predict what would happen in the core of a nuclear reactor when something went wrong with the more mundane parts on which its safe operation depended, such as the control rods or piping systems. His was a difficult job, because the codes had been developed over the course of many years by individuals who had long since left the lab or moved on to management positions, often elsewhere. For all practical purposes, the aptly named codes were enigmatic black boxes into which my office mate fed input and from which he received output that it was his task to interpret in the context of nuclear reactor core physics. Sometimes the results made sense; sometimes they did not. When they did not, it was my office mate's and his colleagues' chore to figure out whether the reactor design was lacking or whether it was modeled improperly within the computer code. This latter situation was not very easy to see, because it is very dark inside a black box. Furthermore, finding the computer code at fault was never the desirable answer, since so much work that had gone before had been validated by means of the code.

Because he was a scientist by training, my office mate wanted to understand what was what. Even if the computer code was an artifact, albeit a complex one that seemed to have taken on a life of its own, it became the object of scientific scrutiny. He agonized over interpreting computer results generated from a model that was supposed to be—but was not—transparent. To him, the questions he was asking the black box were important, potentially questions of life and death, and he wanted to do his job to the best of his ability. He got little sympathy from some of his section colleagues, who generally had become hardened and resigned to the fact that they did not fully understand what they were doing with the digital tool whose inner workings they did not fully understand either. One morning my office mate, who had wanted only to do real and good science, decided never to report for another day of what had come to be for him Sisyphean work. May he rest in peace.

Real scientists, of whom my office mate considered himself one, do not like to compromise their goals of understanding even dark parts of the universe, whether they be black holes or black boxes. Many self-assured scientists, of which my office mate was not one, also like to believe that they are intellectually superior to those who make boxes and paint them black, that the pursuits of scientists are deserving of higher recognition than those of engineers. That is a difficult position to maintain when the job involves feeding one of those black boxes with bits of data and disposing of the output as if cleaning out a litter box. Even if the scientists are not working on problems at the frontiers of science, their role models tend to be scientists who are, and they are thus part of the grand scientific enterprise of probing the universe for eternal truths. Some physicists, in particular, who seek to understand everything from the infinitesimal scale of subatomic particles to the expanding extent of the Big Bang, sometimes appear to think of themselves as special. The physicist Ernest Rutherford, who is credited with the orbital theory of the atom, supposedly once said, "All science is either physics or stamp collecting." Biologists, on the other hand, can be more modest; to Thomas Huxley science was but "organized common sense."[7]

While nuclear physicists tend to be especially prone to hyperbole, they also have demonstrated a heightened sense of playfulness, perhaps prompted by the fact that much of their work depends on inference and metaphor, rather than on direct observation and concrete explanation. Thus, the particle physicist Murray Gell-Mann found inspiration in James Joyce's obscure and difficult novel *Finnegans Wake* for at least the spelling of the word *quark*, one of the posited fundamental constituents of matter. According to Gell-Mann, it was Joyce's enigmatic line "Three quarks for Muster Mark!" that served as the inspiration.[8]

By comparison, the verbal imagination of engineers is generally down-to-earth, if not downright absent, with few if any literary allusions or plays on words in their work. This may be due in part to the fact that the typical focus of engineers is on concrete things of human scale that have roots that reach deep into the soil of civilization. Engineers do not need to imagine the unimaginable; they have to imagine the manageable. They have to come up with achievable new solutions to both new and old problems. This does not necessarily require a new vocabulary; even if an engineer invents a new arrangement of new parts to make a new structure or machine, these things will most likely already have names (beams, girders, columns; gears, rods, pistons; etc.) or will be nameable by analogy with already existing arrangements, parts, and machines. It is the rare engineer indeed who would look to fiction or nonsense verse to name a new arrangement of otherwise familiar things.

A common caricature of the scientist, as depicted in many a cartoon featuring one at work, is of an absentminded professorial type, complete with longish uncombed hair, standing at a blackboard full of mathematical symbols and interminable calculations, holding a piece of chalk as if it were Excalibur, at the ready to cut through some Gordian knot of equations and untie a bound-up secret of the universe. This cartoon scientist wants to look, act, and think like Einstein: casually and comfortably dressed, if not somewhat unkempt and disheveled; unconventional, but in a curiously impish and self-conscious way; irreverent and individualistic, except when it comes to dressing in a nonuniform uniform and championing his specialty.

Another caricature of the scientist has him or her in a white lab coat, often worn over a shirt and tie or proper blouse. Looking not unlike a medical doctor making the rounds—minus the stethoscope—this scientist is often depicted in an animated mode, with hands holding up a test tube or gesturing at some rats in a maze or mice on a treadmill. This is also the image of the scientist-manager, overseer of a small army of worker scientists. As conformist as IBM engineers once were said to be, this scientist, even more so than the hirsute and rumpled kind, is deeply engaged in what has been termed normal science, steeped and enveloped in the paradigm of the time.[9]

Surprisingly, as seriously as they take their quest for truth about the universe, not all scientists may reflect very deeply upon the nature of science itself. According to the astronomer Chris Impey, "Scientists aren't prone to introspection about what they do; they just get on with it." And the computer scientist Ronald C. Arkin, who serves as an advisor to the Army on robotic weapons, holds that it is "historically a pitfall of many scientists" to "have their nose to the bench."[10] As a rule, they appear to read little philosophy of science, but they do know history of science. They are highly conscious of the tradition of their field and the biographies of its founders, and they are very aware of viewing themselves as standing on the shoulders of its giants to see farther into the future of the field. (In fairness, engineers read even less philosophy of engineering—a genre in which few books are even published—and many engineers might be hard-pressed to say upon whose shoulders they stand.)[11]

There is a fictive category of scientist that is curiously depicted by what might be described as Einstein in a lab coat. This is the so-called mad scientist, who is often the alter ego of a brilliant and humanitarian one. The mad scientist always has a practical, if evil, goal: a resuscitated monster, a weapon of mass destruction, a mind-controlling device. Thus, rather than pursuing science for its own sake, the mad scientist merely uses scientific knowledge and know-how to create what never was (and mostly never will or should be in the real world). The mad scientist is a strange hybrid of the scientist and the engineer, presumably with all the intellectual powers of the former and all the supposedly crass practical (even though antisocial) goals of the latter, at least according to the stereotype of each.

However they end up, scientists tend to start young, collecting bugs and rocks, peering through microscopes and telescopes, taking apart all sorts of things natural and artifactual, and truly enjoying basic science courses and experiments in school. Since the words *engineer* and *engineering* are seldom if ever mentioned in elementary and middle school, future engineers and scientists are largely indistinguishable in the classroom or at the science museums that they haunt during summers and on weekends—until peer pressure and hormones lure some away to sports and the opposite sex. The science writer Natalie Angier has observed that by junior high school "science becomes the forbidding province of a small priesthood—and a poorly dressed one at that." She notes further that "adolescent science lovers tend to be fewer in number than they are in tedious nicknames: they are geeks, nerds, eggheads, pointy-heads, brainiacs, lab rats, the recently coined aspies (for Asperger's syndrome); and, hell, why not 'peeps' (pocket protectors) or 'dogs' (duct tape on glasses) or 'losers' (last ones selected for every sport)?" This is a novel perspective, since Angier does not look down on the stereotype of the pocket-protector-wearing engineer but gives him equal billing with the egghead, a term most people associate with the scientist.[12]

Elsewhere in her book, she speaks of troubleshooting a problem with a vacuum cleaner and preparing hollandaise sauce as examples of doing science. Whereas many scientists and science journalists might express surprise if not disdain at such a liberal and nonjudgmental convolution of the pure and applied sciences, as the practical ones that deal with vacuums and cooking are often called, Angier speaks unabashedly of their equivalence. She is, of course, thinking in terms of method, and she is entirely correct.[13]

At the same time, by her own method Angier comes close to propagating the misconception that troubleshooting is tantamount to fixing a broken piece of equipment. In fact, it is a lot more, and for engineers more typically involves getting to the root of a systemic problem. To troubleshoot is to find out and understand why something is not doing what it was supposed to do, whether it be a scientific hypothesis, a mathematical proof, a machine not functioning correctly, or a dish not piquant enough.

Most engineers grow weary of the remark "Here comes an engineer—he'll fix this broken appliance." I experienced this one Fourth of July, when my wife and I were invited to a barbecue supper at our neighbor's house. When we arrived, I went out to the deck, where the fire was being tended by my neighbor standing at a gas grill. As I approached, he greeted me and in the next breath said, "You're an engineer; you should be able to fix this gas tank." It seemed that when he turned the valve handle on the tank, nothing had happened, and so he was starting a charcoal fire the old way, with wadded-up paper. He explained that he had taken the tank to the hardware store and was assured that it was full and that all he had to do was give it a good drop to dislodge the sticky valve. He was reluctant to drop what he conceived of as a potential bomb. Perhaps an engineer would be willing to do it? I should have been flattered that he thought I might fix the problem. I couldn't.

But engineering is not simply about fixing things. When an engineer's automobile develops problems, he, like most other people, will take it to be serviced by an experienced mechanic. Even if he believes that he has diagnosed what is causing the trouble, he is not likely to have the proper tools, parts, or know-how to reach and replace what needs doing. As odd as it might seem that someone who can design an automobile might not be able to fix one, that is the reality of it. Just as a coach might be able to design and diagram a scoring play, so the engineer might be able to come up with an efficacious arrangement of pistons, rods, and gears. But just as the coach and his assistants could not effectively execute the designed play against a formidable opponent on the basketball court, so the engineer could not likely build the machine to his own exacting specifications. There are mental skills and there are physical skills, and while they are usually present in some combination in engineers and mechanics alike, they do not necessarily transfer from one to the other. It is generally easier to move back and forth between engineering and science.

In the post–World War II years, with their emphasis on science at the expense of engineering, a good number of engineers were educated and acted more like scientists than engineers. Still, their focus of study and research was not the natural world but the world of engineered things, and so they came to be known as engineering scientists. It was engineering scientists who gained control of university programs in engineering and propagated themselves through graduate education, but the social demeanor of these and of engineers of all types is generally more subdued than that of scientists. One engineer riddle asks how you can tell an extroverted engineer. The answer is that she is the one who looks at *your* shoes during a conversation with you. And there are numerous self-deprecating jokes that engineers tell about themselves. One of the most familiar has—like jokes of all kinds that get dispersed via an oral tradition—many variations, but generally goes something like this: During the French Revolution, three men were set to be executed by decapitation. The first, a priest, had his head placed on the block beneath the guillotine blade, but it did not fall when the release cord was pulled. By tradition, he was set free, and as he walked down from the platform he proclaimed to have been saved by the grace of God. The second man, a lawyer, was next to have his head placed in the path of the blade. Again, when the cord was pulled, nothing happened. He, too, was set free, and as he walked away he declared that justice always prevails. The third one set to be executed was an engineer. As he ascended the steps toward the guillotine, he looked up at the structure and spotted a wedge of wood blocking the blade's descent. Before placing his neck beneath the blade, he stopped and announced to his executioner, "I see what the problem is, and I can tell you how to fix it."

Generally speaking, the perception of naive engineers being able to fix things has contributed to the confusion between college-educated engineers, whose name derives from the Latin *ingenium*, connoting a talent for devising ingenious devices and systems, and the locomotive-driving kind, who take their name from the engines they operate and tend. It is not that the latter cannot also be ingenious, but they generally are not trained in mathematics and the sciences (including the engineering sciences) and have not learned the formal methodological tools associated with analysis and design. While denim-clad train-driving engineers are expected to focus on the long and narrow right-of-way before them, white-collar engineers are expected to have well-developed peripheral vision and so see also to the sides of the right-of-way and even to project to conditions well beyond the horizon. They are expected not only to react to what might be on the tracks and right beside them, but also to anticipate future conditions and to conceive, design, and bring to reality new configurations of track and the machines that run on them. Mathematics and science have traditionally helped them to do this; now the computer, which often incorporates the math and science in black boxes, does.

Scientists and engineers have a good number of mental abilities and creative skills in common, but that is not to say that all scientists can do engineering, or all engineers science. In addition to knowledge and skill, qualities like affinity for a task and personality come into play. Sometimes, quite frankly, there is also a degree of snobbishness involved in whether people who think they are in a more privileged profession will stoop to do the work that they consider beneath them. Thus, some scientists, even if they could design a piece of equipment for their experiments, would not think of doing so, just as some engineers would not think of trying to fix a piece of equipment they themselves had designed.

Sometimes the distinctions are indeed difficult to draw. In a 1966 book published in the Life Science Library, the engineer was said to typify the twentieth century; without him, "our contemporary life could never have reached its present standard." But at the same time, "despite the essential part the engineer plays in the progress and well-being of humanity, to many he is a blurred figure, his exact role imperfectly understood." The mid-century, male-dominated profession was not sufficiently distinct from science:

> One reason for the hazy impression left by the modern engineer is his close association with the scientist. Both men look alike, talk alike, worry over similar mathematical equations; the guard at the plant gate who checks their identification badges often cannot tell which is which. In fact, in such industries as plastics and communications, it is difficult to determine where the scientist's work ends and the engineer's begins.[14]

Thus it is, and seemingly also as a way of dismissing their individuality, that engineers are often subsumed by careless journalists and laypersons into the general rubric of *scientist*. Engineers are seldom in their own right the subject of magazine cartoons, but they are depicted in newspaper comic strips. *Dilbert* is the most conspicuous: the title character is an engineer who wears short-sleeved shirts and a tie that is said to be so old and out of fashion that it curls up into a permanent flip. He often is depicted with multiple pens and pencils in his shirt pocket. In early 2009, with the economy continuing to weaken, Dilbert had time on his hands at work and so started his own Internet business in his cubicle. When he was found out, he was fired and informed that his business and all of its intellectual property belonged to the company, not to him. He was eventually hired back to fill his old position, but at a much lower salary, and his familiar daily comic-strip activities resumed.[15]

A few years ago, a one-minute video clip about Dilbert was circulated on the Internet, typically as an e-mail attachment. It was often sent with little more introduction than "You should enjoy this." Cryptically titled "The Knack," the video shows a young Dilbert and his mother in a doctor's examination room. As little Dilbert sits beside her on the exam table, his mother explains to the doctor that she is worried about her child because "he's not like other kids." When the doctor asks her to elaborate, she tells of once leaving Dilbert alone for a short period of time only to come back to find that he had disassembled the clock, television, and stereo. When the doctor assures her that such behavior is "perfectly normal," she admits that what really worries her is that Dilbert used the components to make a ham radio set.

The doctor mutters, "Oh, dear," and explains that normally he would run an electroencephalogram on the child but that the necessary device was not working. In the meantime, Dilbert has gotten down from the exam table, opened up the EEG machine, and fixed it. Upon seeing this, the doctor admits that Dilbert's condition is worse than he had feared. He turns to the mother and announces, "I'm afraid your son has . . . the knack." The doctor then explains that "the knack" is a "rare condition, characterized by an extreme intuition about all things mechanical and electrical and utter social ineptitude." When Dilbert's mother asks if her son can have a "normal life," the doctor responds, "No, he'll be an engineer." Upon hearing this, she sobs.[16]

Although the dialogue suggests that the video clip is ridiculing engineers, I got the distinct impression when viewing it that the animated cartoon was not necessarily presenting engineers—or even engineers-to-be—in a negative light. Young Dilbert, who is generally quiet throughout the vignette, comes across as a charming little boy, politely listening to adults talk about him. While his mother explains her concerns, he dangles his legs, like a pair of pendulums, sometimes in and sometimes out of phase with each other, as if enjoying the difference. When the doctor hits Dilbert's right knee to test its reflexes, the child utters a little "oop," and both legs rise. To me, this kind of light touch injected into his personality separates Dilbert from the nerds. He is a real kid with a real sense of humor.

When he gets the inoperative EEG machine to start, Dilbert ironically becomes the hero of the tiny drama. While the doctor and mother had talked, Dilbert fixed what the doctor evidently could not or would not. But there should be no stigma to having the ability to do so and to be destined to become an engineer, since engineers design and understand the workings of the countless medical devices that enable doctors to diagnose and treat all kinds of conditions. Catheters, stents, pacemakers, prostheses, artificial hearts and joints—all these so-called medical devices are in fact products of ingenious engineering design by ingenious engineers.

To the best of my knowledge, I was never diagnosed as having "the knack." As a child, I did enjoy taking things apart, but I could not always put them back in working order, much less make a ham radio out of the parts. I ended up an engineer not because of anything like "the knack" inherent in me. It was the successful launch of the world's first artificial satellite by the Soviet Union that sealed my fate to become an engineer. When Sputnik demonstrated the inferiority of American engineering, there was a concerted national effort to encourage students who had done well in science and mathematics to study engineering in college. My high school guidance counselor looked at my standardized test scores and knew exactly what to advise me to do.

The typical first-year college engineering curriculum of the time was heavy in the science and mathematics courses (chemistry, physics, calculus) that are prerequisite to engineering ones. Those of us who were good in those basic subjects to begin with naturally ate them up. But as we advanced deeper into the engineering curriculum, with its courses relating not to the idealized world of science and mathematics but to the real world of machines and structures, my friends and I began to wonder why, if we were so good in science and math, we were not majoring in those subjects. It did not occur to us at the time that there was very little if any exposure to engineering in high school, unless we took the mechanical drawing, shop, or auto mechanics courses that were intended to be taken by vocational students—not by the college-bound. For all our doubts, however, none of us switched out of engineering into science. Increasingly, we came to realize that engineering, especially as embodied in its central activity of design, was in fact more than math and science. As engineers, we were going to be in a position to change the world—not just study it.

In the last decades of the twentieth century, among the most dramatic and visible contributions of engineers was the enormous growth in numbers and uses of digital computers, especially in conjunction with the Internet and the World Wide Web. The image of constructing an information superhighway was commonly invoked, but some in the industry harked back to a prior technology for a metaphor—they saw themselves "building the railroads of the future." Their concern was not so much the contents of the freight cars that their engines were pushing and pulling but with the great rail yards of countless parallel tracks into which the trains of rolling stock came, were uncoupled, switched, sorted, recoupled, and sent out again. This is the data center that receives and dispenses packets of information via its servers, those electronic engines that push and pull bits and bytes of facts and flame through fiber-optic cable, microwaves, and even twisted pairs of copper wire.[17]

The data center, as many a suite that housed an early mainframe computer was referred to, has always been the heart of any computing network. Today, it is where data is sent, sorted, stored, and dispensed, as if it were money in a bank. Efficient and reliable data centers full of servers are obviously central to the smooth operation of the Net and the Web, not to mention business, commerce, and government. But data centers have not been a very visible or sexy part of high technology.

According to one view, the stars of Silicon Valley have traditionally been the "charismatic marketing visionaries and cool-nerd software wizards," while the mechanical engineers responsible for the design and operation of computer data centers were considered to be "little more than blue-collar workers in the high-tech world." Yet the data centers, packed full of servers and storage devices so essential to the digital economy, were expected to be kept going in spite of rising costs, especially for the energy to run and cool them. With the growth of the Internet, efficient and reliable data centers became increasingly critical to the smooth operation of businesses of all kinds. Given their central role, data center experts can no longer be taken for granted.[18]

Engineers came to be more visible and appreciated within the data center field because of pressing issues of growth and upward-spiraling energy costs. After an almost fivefold expansion over the decade ending in 2007, there were nearly twelve million servers operating in more than six thousand data centers in the United States. Many of these were in need of upgrading, and it was estimated that 30 percent of American corporations were deferring initiatives in new technology because of limitations in their data centers. Among the limitations were insufficient and inefficient air-conditioning systems, which are essential to carry away all the heat generated by the servers. According to one estimate, by 2020 the world's data centers collectively "could well surpass the airline industry as a greenhouse gas polluter." It is ironic that in the world of cutting-edge high technology the mundane problem of keeping a computer room at an appropriate temperature for the machines to work properly came front and center, but it did because air-conditioning could account for half of all energy consumed by a data center.[19]

Just as solving the mundane problems of water supply and waste removal contributed so much to the improvement of our quality of life in the past, so the unglamorous task of keeping computers comfortable provides an essential component to the efficient and reliable running of our present-day technology. Common to such activities, even though a century apart, is the necessity of both a scientific understanding of the problem and an engineering approach to its solution. The same approach is necessary to solve other problems of the twenty-first century, whether they involve macro-issues relating to energy, the environment, and public health or micro-issues relating to unsafe water supplies, localized pollution, and nanotoxins.

4

Which Comes First?

In some dictionaries, engineering is defined simply as "applied science." The implication is that the design of anything made, from airplanes and bridges to yo-yos and zippers, is just a matter of recognizing and assembling the appropriate scientific principles, expressing the relevant scientific laws in mathematical terms, programming the equations into a computer, turning the metaphorical crank, and waiting for the plans to fall out. In this view, design is not unlike dropping a coin into a vending machine, pushing the right buttons, allowing the gears to rotate, and taking the snack or drink out of the tray into which it falls. But where is the science of vending machines? Out of what equations or laws do they arise? To put it as bluntly as Barry Allen, a philosopher who has written about truth, technology, and civilization, has, "It's a joke to say that engineering is applied science when engineers are past masters at taking knowledge where science cannot penetrate."[1]

There is certainly a lot of science to apply in solving many an engineering problem, but that is not the same as saying that engineering derives directly and fully from an application of science. Science is a tool of engineering, and as no one claims that the chisel creates the sculpture, so no one should claim that science makes the rocket. Relying on nothing but scientific knowledge to produce an engineering solution is to invite frustration at best and failure at worst. According to an assessment of the collapse of some condominiums under construction in Florida, "Structural engineering isn't rocket science. Evidently, it is considerably more difficult."[2]

Although it has been said that "applied science was an occupation for second-rate minds," in fact engineering and technology often precede science, because so many instruments and devices are needed to carry out the experiments essential to making scientific observations and testing scientific hypotheses. In Oxfordshire, within a massive bunker known as "the Monolith," there is located a prolific source of neutrons. The machine is called ISIS, after the Egyptian goddess of fertility, and it enables scientists to peer into materials so that their atomic structure can be understood. Such knowledge is said to have contributed to our understanding of everything from a blockage in an oil pipeline to a baby's respiratory system. The device has also helped engineers design ultra-strong and ultra-light airplane wings. But because something helps engineers do something does not mean engineers cannot do without it.[3]

As much as the scientific understanding of the structure of materials gained from ISIS provides scientists and engineers with insights for making new things, the unique machine itself had to be designed without the benefits of the knowledge it has been used to generate. In fact, the design, construction, and operation of ISIS has been described as "an immense engineering feat." It accelerates protons to 84 percent the speed of light by means of electromagnetic fields that are ten thousand times the natural magnetic field at the surface of the Earth. At that speed, in one second a beam of protons could circumnavigate the globe six times. It is when such a beam is directed into the core of the Monolith that secrets of nature can be revealed. But which came first, the engineering or the science?[4]

It is not just in high-energy physics that cutting-edge engineering achievements are essential for scientific progress. The success of the Human Genome Project, in which biological scientists identified and catalogued the twenty-odd thousand genes in human DNA, owed much to the sequencing machine developed by a team led by Leroy Hood, a medical doctor and biochemist. According to Hood, he was greatly influenced by a professor at the California Institute of Technology, who advised him, "If you really want to change a discipline, invent some new technology that will let you go beyond what people have seen before." In the life sciences, as in the physical sciences, technology often leads the way.[5]

This point was emphasized in an editorial by Bruce Alberts, editor in chief of *Science*, on the occasion of the magazine's naming its choices for the ten "major scientific breakthroughs" of 2008. These discoveries ranged in scale from the microscopic to the astronomical, involving advances relating to protons, cells, and genes, as well as to the discovery of an "exoplanet" located one hundred light-years from Earth, thus providing a "breathtaking illustration of the tremendous reach of science." Alberts also pointed out that the year's breakthroughs revealed another important aspect of science, and this was that "new technologies promote its advance." He noted that the scientists credited with the past year's most notable advances employed instruments and techniques that were "unimaginable" when he was a young scientist in the 1960s. The reliance of science on technology that comes from invention and engineering is not always so prominently acknowledged, let alone believed.[6]

I was once asked to give a talk to a group of science journalists who were meeting in my hometown. I decided to talk about the design of bridges, explaining how their form does not derive from a set of equations expressing the laws of physics but rather from the creative mind of the engineer. The first step in designing a bridge is for the engineer to conceive of a form in his mind's eye. This is then translated into words and pictures so that it can be communicated to other engineers on the team and to the client who is commissioning the work. It is only when there is a form to analyze that science can be applied in a mathematical and methodical way. This is not to say that scientific principles might not inform the engineer's conception of a bridge, but more likely they are embedded in the engineer's experience with other, existing bridges upon which the newly conceived bridge is based. The journalists to whom I was speaking were skeptical. Surely science is essential to design, they insisted.

No, it is not. And it is not a chicken-and-egg paradox. The design of engineering structures is a creative process in the same way that paintings and novels are the products of creative minds. Just as there can be no critical analysis of a work of art until it is at least sketched out, so there can be no scientific discussion of a bridge until there is a specific concept of a bridge type laid down. The scientific discussion will differ in detail depending, for example, upon whether the bridge is an arch or a beam. And even though the same fundamental principles of mechanics will apply to both, the application of those principles will proceed differently depending on which geometric form is chosen. The scientific analysis can help engineers choose between a suspension bridge with concrete towers and a steel deck and a cable-stayed bridge with steel towers and a concrete deck, but those distinctions must be made and chosen between by the engineer considering the physics (and economics) of the problem and not by the physics alone.

It has not been science and scientists but science fiction and science-fiction writers that have anticipated some of the greatest engineering achievements of all time. Jules Verne had men reaching the Moon a century before the Apollo 11 mission, and Arthur C. Clarke proposed a system of geosynchronous telecommunications satellites in 1945, two decades before they became a reality. Clarke was also well known for his three "laws of prediction," which declare:

- When a distinguished but elderly scientist states that something is possible, he is almost certainly right. When he states that something is impossible, he is very probably wrong.
- The only way of discovering the limits of the possible is to venture a little way past them into the impossible.
- Any sufficiently advanced technology is indistinguishable from magic.[7]

All three laws might be said to have been applicable to the development of steamships. As late as the mid-1830s, there were no scheduled transatlantic crossings because sailing ships were at the mercy of the winds, and a voyage from England to America could take anywhere from a couple of weeks to a couple of months. The advent of the steamship, in which paddle wheels could propel the craft even when the sails hung limp, promised the possibility of a more predictable crossing time. However, the prevailing "scientific belief" was that ships could not carry enough coal to achieve such a feat economically. According to the contemporary "expert" Dionysius Lardner, no ship could travel the entire distance between Southampton and New York solely under steam generated by the coal it carried on board and still have room for paying passengers. As ships grew larger, the normal science of the time appears to have maintained, so would the power that they needed to propel them and hence the amount of coal that they would have to carry. It thus would have been an impossible quest. The great Victorian engineer Isambard Kingdom Brunel, among others, recognizing that the volume of a ship increased at a greater rate than the resistance it encountered in cutting through the water, believed differently. Thus a larger ship could carry disproportionately more coal to feed its boilers. Brunel's *Great Western* provided a counterexample to Lardner's hypothesis. Brunel went on to design the *Great Eastern*, which proved capable of reaching from England far into the Indian Ocean with the coal carried within its enormous hull.[8]

There are numerous other examples from the history of technology that make it clear that engineering can proceed even in the absence of a complete and correct preexisting scientific understand-

The Victorian engineer Isambard Kingdom Brunel, shown here at the first attempt to launch his Great Eastern, *designed steamships that achieved what scientists said was impossible.*

ing of the natural phenomena involved. The steam engine had its roots in ancient technology, where the pressure of steam generated by heating water in a closed vessel was employed to drive amusing and mysterious devices like animated toys and self-opening doors. In the late seventeenth century, primitive steam engines began to be employed to pump water out of mines, but these were inefficient and slow-acting machines. Early in the next century, Thomas Newcomen devised his "atmospheric engine," which was such an improvement that it was credited with revitalizing the mining industry in north-central England. However, Newcomen's machine was still not very efficient, a deficiency that was addressed in the late eighteenth century by employing the separate condenser devised by James Watt. The steam engine pumped a lot of water out of mines and, when mounted on a wheeled carriage to make a locomotive, covered a lot of railroad track before a science of thermodynamics was fully laid down. Indeed, it was a lack of understanding of the still to be articulated scientific principles being exploited in the steam engine that itself led to the engineering science of thermodynamics. This not only explained the operation of the relatively complex machine but also pointed to areas in which its efficiency could be improved. Since steam engines obviously worked without a full understanding of the science behind them, they did not need thermodynamics to exist: they had clearly been invented, designed, built, and operated well beforehand.[9]

Guglielmo Marconi, who "never let a discouraging theory stand in the way of an experiment," fervently believed that he could send wireless telegraphy signals over hills, mountains, and even the ocean, something that many of the world's leading physicists had declared impossible. Marconi had inferred that the invisible electromagnetic waves discovered in 1887 by Heinrich Hertz could serve as carriers of electrical signals from aerial to aerial without their being physically connected. Marconi began his experiments on a small scale within a single room and progressively increased the distance between transmitter and receiver until he successfully received in St. John's, Newfoundland, a simple Morse code dot-dot-dot sent from Poldhu, located in Cornwall, England. Although Marconi did not claim to understand how they did it, he demonstrated that signals could overcome the apparent barrier imposed by the curvature of the Earth, thus laying the foundation for radio. It was only after Marconi's achievement that the ionosphere off of which the signals bounced was discovered.[10]

The airplane is another classic example of a thing being designed, built, and operated before there was a complete scientific explanation of its working principles. The Wright brothers were neither scientists nor engineers in any formal sense, but that is not to say that they did not apply scientific and engineering methods to develop a machine capable of powered flight. They read what they could find on the problems, experience, and phenomena associated with flight, and they experimented with different designs and profiles for wings and propellers, which they thought of as rather complicated rotating wings, to find the best shape for their purposes. They looked to ship propellers for guidance, but found that "there was no theoretical base" for their design. When a Wright propeller broke in flight, it revealed a weakness that was fixed by a redesign. What the brothers could not appeal to in their quest was a fully predictive engineering science of aerodynamics. The development of that would follow the existence of airplanes whose powered flight begged for analysis. Indeed, airplanes would be flying for decades before there was a full physical and mathematical explanation of why wings worked.[11]

Ironically, in the preface to an augmented edition of Orville Wright's illustrated history, *How We Invented the Airplane*, the Wright brothers are incorrectly described as "true scientists" and their invention as "one of the best documented of all scientific achievements." The brothers did resort to scientific research when it was necessary for them to advance their engineering objective of powered flight, but that is not an uncommon recourse when engineers are confronted with a paucity of correct preexisting knowledge. Indeed, as Orville Wright pointed out in a deposition given in conjunction with a lawsuit claiming that a prior patent had been violated by the Wrights, distinguished contemporary scientists were inclined to believe a flying machine capable of carrying a person would be impossible. The chief engineer of the U.S. Navy claimed that, if such a machine were possible, it would prove to cost more than the most expensive battleship.[12]

As we have seen, rockets were first the toys of amateurs and the stuff of science-fiction writers. There was a lot of seat-of-the-pants experimenting and trial-and-error development. Whatever chemistry was available to mix propellants and whatever physics provided insight into the trajectory of objects in a gravitational field were exploited, but neither chemistry nor physics books had ready-to-use formulas for this new application. There obviously was no mature rocket science before there were rockets; that notoriously difficult engineering science did not lead but followed the design and successful flight of engineered rockets.

The development of the intercontinental ballistic missile provides a further persuasive example. According to Simon Ramo, who as chief engineer of the program has been called the father of the ICBM:

> The project was full of problems. From the outset, we had to consider that the re-entry into the atmosphere might be virtually impossible to solve. Some scientists made calculations that indicated that there was just no conceivable way, within the understanding of the laws of nature, to get rid of the heat on re-entry, and that no material could ever be invented that could stand it. They spoke of the need for making the nose cone out of "unobtainium." Others pointed out that if we were to guide a ballistic missile halfway around the world to a target and hit it with military accuracy, so that the whole project would make sense, we would have to know the velocity of light to one more significant figure. If so, this would require putting some pure scientists to work to measure this basic constant of nature more accurately than it had ever been measured before. If they succeeded, those scientists would deserve a Nobel prize.[13]

Needless to say, ICBMs have existed for over half a century now, but no Nobel Prize has been awarded for a more accurate measurement of the speed of light.

When the mission to the Moon was being systematically designed in the 1960s, there was considerable uncertainty and debate about the nature of the Moon's surface. This naturally presented problems to the engineers responsible for designing a landing module that would set down on an unknown base. The bottoms of the legs of the lunar lander, which would be manually guided down onto the exotic surface, were fitted with what looked somewhat like outside-down hubcaps, thus providing a broad convex surface over which to spread the weight of the vehicle, just in case that was what was necessary. The true nature of the surface was determined only after the landing, and among the most valuable things the astronauts brought back from their historic voyage were Moon rocks, from which geologists back on Earth could learn a great deal about the Moon's "geology." (Even the word for the science of the Earth had to be drafted into service for the study of nature of this neighboring but foreign body.) The Moon landing is a wonderful example of engineering not only preceding science but making it possible.

During the course of debate over the international treaty that effectively banned the use of chlorofluorocarbons worldwide because of their suspected adverse effect on the Earth's ozone layer, a member of the British House of Lords observed that "politics is the art of taking good decisions on insufficient evidence." The same might be said of the art of engineering. Had the inventors and engineers who brought us steam engines, radio broadcasting, airplanes, rockets, and Moon landings felt a need to wait for sufficient scientific evidence and full theoretical understanding before proceeding, then we would likely still be waiting for these engineering accomplishments and achievements.[14]

According to Barry Allen, who has perceived the strong similarities between the creative processes of art and technology in human experience:

Science never tells an engineer how to put anything together. There is never one logical, rational, calculable, supremely functional way to make anything. Any device could have been made differently from different parts and still worked (technically if not economically). That is why not all bridges look the same, and why their differences cannot be dismissed as subjective, or arbitrary social representation. There is no purely technical rationality and no purely functional artifact. The appearance of inevitability in technology (couldn't have been done another way, form follows function, inevitable progress) is an illusion to which contemplative nonpractitioners especially are susceptible.[15]

The engineering of things is "pervaded by choice," something that cannot be easily said about science or even engineering science, to which the natural and made world are givens. Whatever scientists may wish, they cannot credibly propose a theory of motion that does not comport with the facts of the universe. They have no freedom of choice about how the force on an object is related to its acceleration. To propose anything but mass as the factor of proportionality would be to find falsification after falsification whenever the hypothesis was tested. [16]

The inventor or engineer, on the other hand, can proceed to design machines in ignorance of the laws of motion (but not in violation of them). These machines will either be successful or not, depending on how knowledgeable and insightful (and lucky) the engineer might be about what has not worked in the past and what has almost worked. These "experiments" need not have been the engineer's own: he is free to exploit the work of others as if it were his own. If he strikes upon a configuration of parts that pumps water or propels an airplane or guides a rocket, he can say that he has found *a* solution to the problem on which he was working. But an engineer who understands engineering (and English) will never claim to have found *the* solution.

This is why there are so many different-looking airplanes and automobiles and why they operate differently. Sometimes the differences are intended to avoid patent infringement or make a fashion statement, but mostly they are simply one engineer's solution to a problem that has no unique solution. Early automobiles were modeled after bicycles, some of which by the late nineteenth century had, for "more sedate cyclists," three and four wheels (in spite of what they were called). The stability provided by the extra wheels was welcome when a relatively heavy engine was added to make a horseless carriage. Decades later, the iconoclastic self-taught architect, engineer, and inventor Buckminster Fuller designed his Dymaxion Vehicle as a three-wheeler, which made it able to turn on dime, a very handy quality when trying to maneuver in a narrow alley or park in a tight space. Even the name of his car defied convention, being the inventor's "favorite neologism." He used it to qualify not only his vehicle but also his Dymaxion House, Dymaxion Bathroom, Dymaxion Chronofile (described as a "systematic record of his life, including everything from his correspondence to his dry-cleaning bills"), and other Fuller creations that relied on neither scientific theory nor literary precedent. Ironically, the word *dymaxion* itself did not spring from Fuller's own mind but from that of a consultant hired by Marshall Field's department store, which in 1929 sought a "catchy label" to identify its display of Fuller's scale model of his hexagonal sheet-metal home. The residence was designed to hang from a mast that contained all the electrical wiring and plumbing needed to service the annular structure. The consultant's meaningless portmanteau—a combination of parts of the words *dynamic, maximum,* and *ion*—so appealed to Fuller that he "adopted it as a sort of brand name."[17]

Putting parts of things together to make new things is also the essence of much inventive engineering. Sometimes the original functions of the parts are preserved and amplified in the new object, but sometimes an entirely different form and function results from the grafting. Few engineers are as idiosyncratic in their assemblages as Fuller or the Wright brothers, but all are to a degree arbitrary in how they conceive and execute their designs.

Think of the situation encountered en route to the Moon during the Apollo 13 mission. When one of the spacecraft's oxygen tanks exploded and two of its fuel cells stopped functioning, the situation looked grim for the three astronauts on board. It was not at all clear that the aborted mission would end with the astronauts returning safely to Earth—unless they could fix the carbon dioxide filter on board. As depicted so dramatically in the movie about the event, engineers at Mission Control in Houston assessed what could be cannibalized from the cabin or was otherwise available to be used for repairs on board the threatened spaceship. The items were collected in a cardboard box and then dumped on a table, with instructions to the Space Center engineers to make do with what was there to design a fix that the astronauts could effect in time to save the mission and their lives. What the engineers came up with and the astronauts implemented was a jury-rigged carbon filter that worked well enough to bring the crew home safely.[18]

Most engineering problems are neither encountered nor solved so dramatically as the emergency on Apollo 13, but engineers often do have to make do with what is available in supplier catalogues or on a machine shop's shelves to come up with a solution to a pressing problem that might arise in a factory production line or on a ship at sea. When time is of the essence, there is not the luxury of a prolonged research and development program to produce some elegant solution. While today engineers might use the telephone and the Internet to identify and procure uniquely suited parts, they will be doing what engineers have done for centuries—coming up with a workable solution to a problem, even if in a quick-and-dirty way. Later there may be time to refine it and make it look prettier, but pretty is seldom essential to an immediate need. That is not to say that elegant solutions are any less appealing to engineers than they are to mathematicians.

Mathematicians like to state and prove existence and uniqueness theorems, which assure beforehand that there is one and only one solution to a particular class of problems. That is of great benefit to mathematical physicists who use equations to model the universe large and small, but it is seldom something that engineers working on down-to-earth problems have the luxury of exploiting. Just as there is more than one way to skin a cat, so there is more than one way to solve an engineering problem. It is why we can go to the superstore and find a multitude of different kinds of toaster ovens, washers, dryers, refrigerators, and other small and large appliances. Some of the differences, such as where a door is located and how it opens, may be superficial, but other distinctions, such as the nature of heating elements, or method of agitation, or compressor design and operation, can make the difference between satisfactory and unsatisfactory performance. We can observe the same phenomenon in different brands of laptop computers: the location of their various ports is seldom as important to performance as the kind of central processor they contain.

Digital computers were developed before there was computer science, at least by that name. The first computers were designed to speed up the calculation of trajectories of the likes of rockets, missiles, and artillery shells, tasks that previously had to be done by hand, often under the pressure of time constraints. Such pre-digital computer calculations were often done by teams of people themselves called "computers," who worked in an assembly-line kind of way to turn a series of inputs into a series of outputs that provided settings for launch. Mechanical desk calculators speeded up the process somewhat; the physicist Richard Feynman supposedly once said that "working desk calculators were as important to the Manhattan Project as the fundamental physics."[19] But the advent of the digital computer was revolutionary. Because of their roots in the calculation of trajectories, the earliest computers tailor-made to do so were thought to be such specialized machines that they would have few other practical uses. No one at the time seems to have foreseen the general-purpose laptop—or the Internet.

The Internet as we know it today has its roots in the government-sponsored ARPANET, which was launched in late 1969 but still had fewer than a couple dozen nodes in the spring of 1972. What made the system work was the invention of packet switching, a method of communication whereby messages are subdivided into parts known as packets that are routed individually over a shared network and recombined at their destination. For such a system to develop for worldwide use, a network interface standard was necessary, and this was to be worked out at international meetings. A participant in these meetings reported encountering some unexpected opposition from a foreign delegate. After acknowledging that the functioning ARPANET was "an interesting experiment," the skeptic continued,

> But there is no formal mathematics underlying packet switching technology. You can't mathematically prove that sending individual datagrams via alternate paths works. If there is no math, then there is no science—and if there is no science there can be

no technology. We do not standardize things that are not a true technology.[20]

While it may be the rare scientist who refuses to believe his or her eyes, it is also the rare engineer who realizes all the implications of his invention. Thus, not everyone saw the ARPANET developing into the Internet and World Wide Web, just as most engineers do not realize the ultimate uses of the things they design and develop. The Wright brothers were interested in their machine as a means of demonstrating the possibility of powered and controlled flight, not as the initial step toward jumbo jets that carry packages for next-day delivery to much of the world. In this regard, engineering can be as much of an assault on the frontiers of knowledge as is science. The solutions to problems that engineers come up with today can prove to be the jumping-off points for solutions to the as yet unspecified problems of tomorrow. Of course, new solutions to old problems can also bring new problems. Just as inventions can appear to take on a life of their own and evolve into forms undreamed of by their original inventors and engineers, so can they be full of surprises for their future users.

5

Einstein the Inventor

Anyone can be an inventor and receive a patent. All it takes is a good idea that no one else has yet thought of, or at least that no one else has described in a patent application. Successful inventors have come from all walks of life, and have ranged from obscure autodidacts to world-acclaimed geniuses. Those with considerable grounding in science and engineering naturally have that base of knowledge and experience to draw upon, but those with little formal training can be just as effective as inventors.

The movie-screen goddess Hedy Lamarr, whose first husband (of an eventual six) was involved in the munitions and aircraft business and had a special interest in control systems, was co-inventor of a radio guidance system for torpedoes that was patented under the title "Secret Communication System." Her co-inventor, George Antheil, was an avant-garde composer who scored what have been described as "rhythmically propulsive pieces" for player pianos, airplane propellers, bells, sirens, and the like. In 1940, he and Lamarr were neighbors in Hollywood, where he was living while working as a film composer. She confided in him that she was thinking of leaving the movies and working for the newly formed National Inventors Council, which was established to give ordinary citizens a means of communicating their ideas to help the war effort. The council was chaired by Charles F. Kettering, who was director of research at General Motors.[1]

As her conversation with Antheil continued, Lamarr revealed her idea about controlling a torpedo by changing the frequency of radio signals guiding it, if the transmitter and receiver frequencies could be properly synchronized. Antheil saw an analogy with a technique he conceived of to synchronize sixteen player pianos in his composition *Ballet Mécanique*, which has been described as a "jackhammer of a piece." They sent their idea to the Inventors Council, and Kettering encouraged them to develop the concept further so that they could apply for a patent. They consulted a professor of electrical engineering at Caltech about some details and filed for a patent in 1941. The invention, often attributed to Lamarr alone, was announced in an item in *The New York Times*, which reported that her discovery was so vital to national defense that the government did not allow details of it to be published at the time. The patent was granted about a year later.[2]

Another entertainer-inventor was the ventriloquist Paul Winchell, who shared the stage with his dummies Jerry Mahoney and Knucklehead Smiff. Among his inventions, for which he received over thirty patents, were a flameless cigarette lighter, battery-heated gloves, a retractable fountain pen, an invisible garter belt, and a portable blood plasma defroster. He also conceived of the disposable razor, but decided against pursuing it because he was convinced that people would not buy a product that they would use and then throw away. His most significant invention was an artificial heart, and he received the first patent for such a device.

His early success as a ventriloquist resulted in his postponing higher education until he was in his mid-thirties. Then he became a premed student at Columbia University, studied hypnotherapy, and received the degree of Doctor of Acupuncture from the Acupuncture Research College of Los Angeles. After seeing a patient die on the operating table, Winchell had the idea that an artificial heart could have maintained vital functions during surgery. He shared his idea with his friend Henry Heimlich, the doctor who developed the anti-choking maneuver, and Heimlich encouraged Winchell to build a model and offered help if needed. Winchell's artificial heart was patented in 1963, preempting researchers at the University of Utah, who had been working on a similar device. He eventually donated his patent to the university, and his concept influenced the development of the Jarvik heart, the first to be implanted in a patient.[3]

At about the time that Winchell was patenting his artificial heart, the provost of Stanford University was writing that the contemporary technological world was "simply too complex for untutored genius, however great, to make many significant contributions." But when new technological trails are blazed, there are few if any guides to lead the way. Who were the tutors of Bill Gates and Steve Jobs in the uncharted frontiers of the personal computer industry? The field may be complex today, but it was a virtually blank slate for such pioneers.[4]

Each new patent is akin to a new hypothesis—an assertion that the thing described is novel, unobvious, and will work. If we take the framing and testing of hypotheses as the defining characteristics of science, then any new design (whether patented or not) is something to be tested by the scientific method. Proposing a bold new bridge design, perhaps one that represents the longest suspension bridge theretofore achieved, lays down the hypothesis that the design will in fact work as a bridge. Like any scientific hypothesis, it cannot be absolutely proven; it can only be disproven—or falsified—typically by a counterexample. In engineering, a counterexample often takes the form of a physical failure. Thus, the scientifically rigorous way to test the hypothesis that a new bridge design will work is to build it and let it function as designed. Although the first and usually deliberately heavy load that such a structure supports is typically described as a "proof test," it is in fact only a confirmation of the hypothesis, not a proof in any rigorous mathematical sense.

Not every engineering hypothesis need be subjected to a full-scale, real-time test. It is the nature and power of modern engineering to be able to predict behavior using the calculator and computer. This is what engineers do when they calculate the voltage drop across the capacitors and resistors of a proposed circuit or the forces carried by the beams and columns of a hypothetical skyscraper. Because such components are familiar (preexisting) building blocks of engineered systems, a great deal of engineering science about them has been developed (usually by engineers playing the role of scientists). The calculations of voltage and force that engineers make, when compared with failure criteria, enable the hypothesis to be tested in the design office in theory before the system is built in the field in practice.

Given the almost seamless ease with which engineers and scientists can move back and forth between engineering and science, it should come as little surprise that the so-called scientific method and engineering method are very difficult to distinguish from each other.

In recent years, the U.S. National Academy of Engineering has elected about sixty-five new members annually, and in a typical class approximately one in four does not have a single engineering degree. These members received their undergraduate and graduate education in such fields as chemistry, physics, mathematics, and computer science, but they have accomplished outstanding achievements in engineering. They have, in many cases, worked in industry, making significant contributions to manufacturing and communications. (It should go without saying, but bears saying, that a degree in science is not a prerequisite to becoming a member of the National Academy of Sciences.) Indeed, neither a scientist nor an engineer need ever fully cease being the one in order to be the other. Steve Wozniak, cofounder of Apple and a member of the National Academy of Engineering, has been referred to as "one of Silicon Valley's most creative engineers." In early 2009, he joined the staff of the start-up company Fusion-io, which specializes in providing computers super-fast access to stored data, with the job title of "chief scientist."[5]

If scientists and engineers each can be doing work that characterizes the other, then do they ever really do pure science or engineering? It would seem not, even though the conventional wisdom and way of speaking is contrary to this view. Perhaps the times when scientists are most doing science is when they are testing their (engineered) hypotheses; engineers are most doing engineering when they are involved with designing things that are to be constructed and manufactured, the processes by which engineering hypotheses of all kinds are in fact realized so that they can be tested.

No less a scientist than Einstein had numerous forays into engineering design. On one occasion, he was motivated by the desire to confirm experimentally his theory of Brownian motion, which deals with particles suspended in a liquid or gas. In order to measure the "immeasurably" small voltages expected, he devised an "electrostatic induction machine" that would amplify the quantities to the point where they could be measured by a conventional electrometer. Einstein referred to this invention as his "little machine" (*Maschinchen*).

He first conceived it around 1903, but he himself evidently did not possess sufficient mechanical skill (or interest or desire) to build a satisfactory prototype. This led to a quest for collaborators who would help bring the device to acceptable operational status, a process that culminated in a successful test in 1910 and commercial production in 1912. But this was not the scientist's only excursion into engineering.[6]

According to the historian of technology Thomas Hughes, "Einstein had more than a passing and trivial involvement with patents and inventions," which may have been influenced by his childhood experience, including proximity to the electrical machinery factory operated by his father and his uncle, who held patents for arc lamps and other electrical devices. Even the failure of the family business to compete successfully with larger German manufacturers must have made an impression on the young Einstein, perhaps teaching him the power and value of intellectual property. When their business failed in Munich, the Einstein brothers moved operations to Pavia, Italy, where a fifteen-year-old Albert often became involved in practical problem solving. According to his uncle, "Where I and my assistant engineer have racked our brains for days, this young fellow comes along and solves the whole business in a mere quarter-hour. He'll go far one day."[7]

Young Einstein may have been expected to become a successful electrical engineer. But in 1896 he began studying mathematical physics at the Swiss Federal Polytechnic School, a teacher training college that was widely respected for its engineering and science curriculum. Zurich Polytechnic, as Einstein called it, had an excellent new electrical engineering laboratory, where the young student was exposed to the state of the art in electrical measuring instruments and to the physical principles underlying them. After graduating, though not easily, in 1900, he sought a job as a teaching assistant so that he might continue his studies, but he encountered difficulty doing so and finally accepted a nonacademic position, however not in engineering.[8]

It is well known that Einstein worked in the Swiss patent office from 1902 to 1909. During that period he would have examined many patent applications in the rapidly developing field of electrical devices, including ones having to do with phenomena associated with electromagnetic waves. Understanding how such phenomena were exploited in practical devices demanded the development of a keen sense of conceptualization and visualization. Hughes believes that reading the patent applications of such contemporaries as Nikola Tesla, Lee De Forest, and Michael Pupin (all of whom sought international patents) would have been stimulating to his imagination in important ways. Einstein himself acknowledged that the formulation of patent statements not only presented him with the occasion to think about physics but also assisted him in achieving a means of precise expression. Reflecting upon his experience, he wrote that "a practical profession is a salvation for a man of my type; an academic career compels a young man to scientific production, and only strong characters can resist the temptation of superficial analysis." In other words, as a patent examiner, he felt no pressure to publish in quantity, so he was free to write fewer papers but with substantial content. Also, according to Hughes, for Einstein the "reading and analysis of patent claims and drawings seems to have cultivated not only his visual imagination but also his knowledge of the physical principles that underlay inventors' practical devices."[9]

Einstein became and remained so familiar and comfortable with the patent system that more than a decade after he left his position as examiner he continued to work as a paid consultant to individual inventors and corporations on matters of patents. He was asked to serve as an expert witness in patent cases relating to wireless telegraphy. His association with inventors and inventions appears also to have informed his science. Again according to Hughes, given a more mature Einstein's "use of physical metaphors in his scientific writing, his preference for visual thinking, and his familiarity with inventions, it is safe to assume that he fully recognized the similarity between the intellectual activity of the creative scientist and that of the creative inventor."[10]

Einstein was not only an examiner and advocate for patents; he also became the applicant for and the recipient of patents. Among the countries issuing dozens of patents to him and his co-inventors for such things as refrigerators, pumps, and a "light intensity self-adjusting camera" would be Switzerland, Germany, France, Great

Britain, Hungary, and the United States. His work on practical devices generally took place after he won the Nobel Prize for work on the photoelectric effect. It was Einstein who suggested to the inventor Hermann Anschütz-Kaempfe that he mount his gyrocompass in an armature, thereby making it an alternative to the magnetic compass that came to be widely used for navigation.[11]

Einstein's most frequent collaborator in invention was Leó Szilárd, the son of a civil engineer. Born in 1898 in what was then Austria-Hungary, the younger Szilárd began studying engineering at Budapest Technical University in 1916, but the following year his education was interrupted by military service. After the war, he resumed his engineering studies in Budapest but soon left his native country to escape an oppressive regime. He continued his education at the Technische Hochschule in Berlin-Charlottenburg, but then in 1920 switched to studying physics as a graduate student at the University of Berlin, where he met Einstein, who was by then a professor there.[12]

The older physicist became aware of Szilárd's work in extending classical thermodynamics to systems that fluctuated over time. Although Einstein was at first skeptical of Szilárd's approach, his results so impressed the distinguished scientist that he praised the dissertation that led to Szilárd's receiving his doctorate in physics in 1922. The friendship and mentorship that developed between the two physicists enabled Einstein to advise Szilárd candidly to work in a patent office after graduation, telling him that his own years as a patent examiner were his "best time of all," referring to the lack of pressure on him then to come up with earthshaking physics results. But Szilárd declined the advice, choosing instead to continue working at the university. He maintained his interest in thermodynamics, publishing a landmark analysis of the problem of Maxwell's Demon— which presumably violated the Second Law of Thermodynamics by separating molecules according to their speed and thus was capable of operating a perpetual motion machine. In the same year Szilárd filed a patent application for a cyclotron, which is essentially a large machine for making small particles go extremely fast. Szilárd, like a lot of scientists and engineers, was capable of working on scientific and engineering problems virtually simultaneously.[13]

The genesis of the Einstein-Szilárd collaboration as inventors has been the subject of some speculation. It may have begun when Einstein read a newspaper article describing how all the members of one family died in their sleep from inhaling the poisonous gases that leaked from a refrigerator pump. Thinking like a true inventor, Einstein said to Szilárd, whom he visited with frequently, that there must be a better way to make a refrigerator. One Einstein biographer, Abraham Pais, believed that the two scientists were driven by the simple desire to have a noiseless refrigerator. In any case, the scientist-inventors easily identified the mechanical pump as the weak link in the system. If they could eliminate moving parts, they believed they could eliminate leaks and noise, and they were confident that they could do so by exploiting the principles of thermodynamics.[14]

Einstein and Szilárd did not enter into their engineering collaboration casually or without contemplating the potential value of the patents they anticipated receiving. The Nobel laureate understood that the aspiring academic Szilárd, who was about to give up a salaried position as university assistant to take the unsalaried but more prestigious position of *Privatdozent* (a licensed lecturer who was dependent upon student fees), would have an uncertain income until he advanced in the ranks, and so the two reached an agreement that any invention relating to refrigeration that either of them made would be considered joint property. However, according to their written agreement, Szilárd would receive first any profits from their patents up to the level of the salary of a university assistant, and only afterward would he and Einstein share in them equally. It has been suggested that Einstein was not as interested as Szilárd was in inventing and patenting. Perhaps this was the case, but there is also evidence that Einstein engaged in invention simply because he greatly enjoyed the challenge and pleasure it provided. Whatever their individual motivations, the pair sought wide international patent protection for their refrigeration concept.[15]

In a conventional refrigerator, a mechanical compressor causes the gas being used as the refrigerant to be liquefied under pressure, releasing excess heat. Then, the refrigerant is allowed to expand into a gas, a process that is accompanied by cooling. The cool gas can then absorb heat from inside the refrigerator cabinet, thus keeping its contents at the desired temperature. An alternative means of driving the cooling cycle was to liquefy the gas not through the means of a mechanical piston but through means of heat, such as that produced by burning natural gas. One version of this "diffusion absorption heat pump" operates on an ammonia-water-hydrogen cycle, with the three working fluids being maintained at essentially the same pressure. An external source of heat drives the ammonia—the refrigerant—from an ammonia-water mixture, and the separated ammonia then condenses and flows into an evaporator, where it comes in contact with gaseous hydrogen, causing it to evaporate and absorb heat from whatever is to be refrigerated. The mixture of ammonia-hydrogen vapor then flows into an absorber, where the ammonia is absorbed by water and the hydrogen returns to the evaporator. The ammonia-water mixture is then available to begin the cycle again. A refrigerator employing such a scheme had been patented by a pair of Swedish inventors, Baltzar Carl von Platen and Carl Georg Munters, and was being manufactured and sold in the United States in the 1920s by the Swedish company Electrolux. However, the refrigerator received bad publicity following a fatal natural gas leak in Chicago, thereby giving the market over to mechanical refrigerators.[16]

It was for the von Platen–Munters absorption refrigerator that the physicist-inventor partners Einstein and Szilárd came up with an improvement—utilizing butane, as a refrigerant, in combination with ammonia and water, each fluid having a cycle of its own. Following the 1926 receipt of a German patent, they sought English-language patents, which were issued between 1928 and 1936. Rights to the American patent were assigned to the Electrolux Servel Corporation, but the timing was obviously inopportune, given the tragedy in Chicago.[17]

A prototype of the Einstein-Szilárd refrigerator was built in 1931 by the German company AEG—a European counterpart of the American giant GE—but the machine was never manufactured or marketed, due at least in part to the fact that its development coincided with the Great Depression. Szilárd also believed that neither the pumpless model nor refrigerators generally would have much of a market in Germany. In the meantime, noise and leakage problems

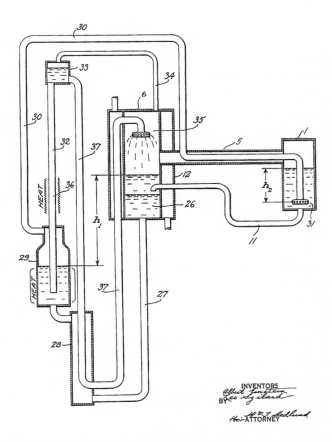

Nov. 11, 1930.   A. EINSTEIN ET AL   1,781,541

REFRIGERATION

Filed Dec. 16, 1927

In 1930, Albert Einstein and Leó Szilárd were granted a U.S. patent for an improved refrigeration system, which had no moving parts. The two scientists were also committed inventors.

with mechanical models had become greatly reduced, thereby diminishing the advantage of the motorless kind. It has been claimed that AEG abandoned the Einstein-Szilárd refrigerator because the prototype was too noisy, but the company may also have abandoned it for other reasons. After all, Einstein and Szilárd had sold to them the patent rights, and so the company could suppress production of the

competing design. Among the most likely reasons that the nonmechanical refrigerator did not displace the mechanical kind was that in
1930 the gas Freon was discovered. Since this was believed to be a
nontoxic refrigerant, the danger of a leaking refrigerating line was no
longer a matter of life and death—the way a leaking natural gas line
threatened to be.[18]

Among the devices that the scientist-inventors had developed for
versions of their nonmechanical refrigerator was an electromagnetic
pump. Szilárd later believed this to be potentially useful in the
Manhattan Project, when it became necessary to choose a cooling
medium for a nuclear reactor. He proposed using the electromagnetic pump to drive a liquid metal through the system. His choice for
the metal was bismuth, which had the desired thermal properties.
The idea was not adopted at the time, but similar technology was
employed decades later in the nuclear industry, when liquid sodium
was used to cool fast breeder reactors.

Einstein and Szilárd applied their knowledge of thermodynamics
to invent their refrigerator, but relevant science isn't always available
or known to inventors and engineers. This is not necessarily an impediment to invention, as so many historical examples have demonstrated. In 1996, David McHenry, an inventor from Stone Mountain,
Georgia, patented a device for making neat bundles out of a mess of
unruly computer cables and has marketed it under the name Kurly
Tie. Key components of the tie are lengths of plastic-coated coiled
cord similar to the kind that connect the handset to the base of an old
telephone set. In his patent, McHenry admitted that he did not know
the "scientific reason" that the familiar coiled cord "does what it
does" in keeping a bundle of cables firmly collected, but "whatever
the reason it behaves as it does it is extraordinarily well adapted to
the purpose." Inventors do not have to know why their inventions
work, but those who are engineers and scientists usually do, even if
they cannot decide or articulate whether they are doing science or
engineering.[19]

Whether a scientist works as an inventor or engineer for humanitarian or pecuniary reasons, or an engineer becomes sidetracked by
scientific experiments to fill in some nonexistent but essential knowledge to complete a design, the practices of science and engineering

become convoluted and intertwined like an uncontained set of wires and cables. Science and engineering are—and always have been—coequal partners in the development of the world of thought and things that help define civilization and culture. When the world is threatened by asteroids, ozone holes, and other potentially catastrophic natural disasters, the ability of scientists and engineers to move freely and beneficially back and forth between engineering and science is of indispensable benefit to the planet. No matter who works on them, the highly challenging and often unprecedented problems presented by such potentially earthshaking phenomena make it highly unlikely that solutions, no matter how technically elegant, will be easy or straightforward to implement. After all, even Einstein and Szilárd could not get an improved refrigerator manufactured. And when things *do* go ahead, there are often unexpected obstacles encountered on the road to success.

6

Speed Bumps

Unanticipated developments can sometimes render ineffective or uncompetitive the most promising invention or design—no matter how brilliant its inventor or how outstanding its designer. As we have seen, this was the case with the motorless refrigerator of Einstein and Szilárd: as long as conventional appliances could leak poisonous fumes, there was interest in the leakless alternative—until nontoxic Freon became available. It was then the standard refrigerant for decades—until the chlorofluorocarbon's effects on the environment became known. But as long as alternatives to Freon were available, there was no need to scrap existing refrigerator designs, and so the Einstein-Szilárd system did not get a real second chance.

All inventions and designs can potentially have active and passive obstacles thrown in their way, thus effectively blocking the road to successful implementation. Moreover, successful designs can also be sidetracked or turned into embarrassing failures because of their inherent qualities. Design aspects that were deliberately introduced to solve a particular problem or to achieve a particular effect can turn out to have undesirable consequences. This phenomenon can manifest itself even in the simplest of things found in the most common of settings.

Energy-efficient compact fluorescent lightbulbs promised to be an environmentally friendly and overall-cost-conscious alternative to energy-hungry incandescent lightbulbs, also known as filament bulbs. In 2007, Australia instituted a ban on filament bulbs, to be effective in 2010. A year later, the energy ministers of the twenty-seven member states of the European Union agreed also to ban the conventional bulbs by 2010. The resultant lower energy consumption was expected to reduce the amount of CO_2 released into the atmosphere by thirty million tons annually. In the United States, the Energy Independence and Security Act of 2007 effectively put in place a ban on filament bulbs by requiring that more efficient lightbulbs be phased in between 2012 and 2014.[1]

Although the compact fluorescent bulb has its roots in the nineteenth century, the modern version was developed by General Electric engineer Edward Hammer during the oil crisis of the mid-1970s. However, because GE did not want to invest in new facilities, the company did not pursue the idea to mass production at the time, and eventually the energy-saving bulb was produced by others. In the 1990s, hotels were among the first establishments to embrace the new bulbs as a means of demonstrating a commitment to the environment while at the same time enjoying sharply reduced power bills. Domestic use of compact fluorescents was slower to take hold, in part because of the harsh light cast by early designs and their much higher purchase cost. However, active promotion by large consumer outlets like Home Depot and Wal-Mart made it clear that over time the savings in utility bills more than paid for the initial investment. As sales and use of the compact fluorescents began to climb, though, a downside to the bulbs attracted attention. Since there is a small amount of mercury (in vapor form) in each compact fluorescent, large-scale disposal of spent bulbs can create significant environmental hazards. To counter this bad publicity, Home Depot instituted a nationwide recycling program and Wal-Mart sponsored "take-back events," where consumers could bring in their blown compact fluorescent bulbs to be disposed of responsibly.[2]

One way to reduce the risk of mercury poisoning has been proposed by Robert H. Hurt, an engineering professor at Brown University. He and Natalie Johnson, a student, in collaboration with others, came up with a material that binds with any mercury that might escape from a broken fluorescent tube. After experimenting with a variety of elements, including sulfur, copper, and nickel, the engineering team found that nanoparticles of selenium worked best. The engineers believe that cloths incorporating the "nanoselenium" could be used to soak up any spilled mercury. Not every technological downside is so easily resolved. Potentially damaging side effects of new technologies can appear to present insurmountable problems for their adoption, but engineers thrive on the challenge for a new or improved invention that a potential failure provides. Failure drives invention.[3]

In the meantime, light-emitting diodes, which for some time had been used in indicator bulbs on electronic equipment of all kinds, became the most efficient of all available lighting sources. But consumers tend to choose short-term price economy over energy efficiency, and so the much more expensive LED, even though it was longer lasting, was not competitive with either the incandescent or the compact fluorescent bulb. (At the end of 2008, an LED replacement for a compact fluorescent cost thirty times as much.) The strategy of the LED industry was first to market its product to large commercial customers, like office buildings and retail stores that burn their lights almost constantly. Such users could save on lighting expenses not only by reducing their electric bills but also by not incurring the labor expense associated with frequently changing conventional lightbulbs, some of which are in hard-to-reach locations. Furthermore, because LEDs do not have the environmental problems associated with compact fluorescent bulbs, it was believed that eventually the LED would displace the compact fluorescent—long before it fully replaced the incandescent bulb. In anticipation of this, Philips Lighting, which had been among the leading makers of compact fluorescents, switched its research and development efforts from those bulbs to LEDs. One technology's bump in the road can be another's takeoff ramp.[4]

But there were technological features of the LED bulbs that potentially put them at a disadvantage when competing against even the long-established incandescents. When used in recessed ceiling lighting, which had become increasingly popular in contemporary homes, the design of the incandescent was such that not only its light but also its heat was reflected downward and so out of the can in which it was housed. In LED bulbs, however, the heat they generated was concentrated near the socket and thus directed up into the recessed fixture, possibly presenting a fire hazard. In order to deal with this undesirable feature, the bulbs were designed with a florid arrangement of fins encircling them to conduct the heat in a more benign direction. This gave the bulbs an unconventional look. As long as the bulbs were concealed within a can or behind a lamp shade, this would not be especially objectionable, but when the bulbs were used in track lighting, where they are exposed, manufacturers were challenged to come up with new approaches. One company chose to emphasize the cooling fins as a design feature, but only their performance in the marketplace will determine whether this was a good or bad bet.[5]

It is unlikely that new technologies introduced to deal with some of the global issues facing planet Earth will ultimately fare any better than compact fluorescent lightbulbs. Solutions that initially may seem to be very promising alternatives to long-established practices that have been found to be problematic—at least from an energy consumption or pollution perspective—can prove to have their own shortcomings. Whether viewed as bumps in the road or potholes in the street, they cry out for leveling out or filling in, which is something that inventors and engineers strive to do.

Dover Road is a gradually rising and falling and gently curving street that winds through the heart of my quiet neighborhood. Although it is by no means a thoroughfare, this narrow, undivided two-lane road is one of the few ways in and out of the 1920s-era subdivision, and so it bears considerably more traffic than the byways.

There was a time when Dover Road was as softly undulating and smooth, and as unobstructed by stop signs, as a lonely country way. It was possible to drive the mile or so between Old Chapel Hill Road and Hope Valley Road at an uninterrupted pace that seemed just about right for the place and time. It is hard to think of cars traveling too fast, because the curves and dips and crests would surely have mitigated any excesses. Nevertheless, some must have occurred or were imagined to have occurred. For whatever reason, a few years ago abrupt obstacles began deliberately to be installed in the road. The "speed humps" appeared without warning, and now there are a total of six of them, heralded by signs at either end of Dover warning SPEED HUMPS AHEAD. These deceptively simple interruptors of traffic are also known as "road humps," "speed tables," and "sleeping policemen."[6]

Designing a speed hump might not seem to be a very challenging engineering problem. After all, it is just a long mound of asphalt laid down across the road to make drivers reduce speed lest they ruin their car's shock absorbers or bump their heads on its ceiling. But the shape of that mound is crucial, and we have all observed that different kinds of speed humps have different effects on different vehicles moving at different speeds. One morning, elsewhere in town, I observed a road crew installing a new speed hump of built-up asphalt, and that evening I saw it being removed. Perhaps the hump was so difficult to negotiate by any car at any speed that complaints were immediate. After a few days, the hump's profile was redesigned and rebuilt in such a way that I now can drive across it at the posted speed limit without feeling either an uncomfortable rise or a sudden fall of my car and my stomach. This is the way speed humps are supposed to work.

Although the terms are often used interchangeably, the difference between a speed hump and a speed bump in America appears to be the extent to which they reach in the direction of travel. A bump is relatively narrow, so almost as soon as a car's wheels climb up on it they fall down on the other side, usually at a rate faster than the shock absorbers can respond. Such a bump is typically encountered in parking lots, where the desired speed limit might be as low as five miles per hour. The only way to negotiate these bumps without a jarring effect is to move at a snail's pace. They thus accomplish the desired objective.

Speed humps are elongated in the direction of travel, sometimes being just gently rising and falling mounds and sometimes incorporating a flat crosswalk that forms a plateau between the bump halves. Such humps are typically installed on streets and roads that have speed limits in the range of fifteen to twenty-five miles per hour and where there are likely to be pedestrians. When properly designed, the humps work like a charm; when not, they are scarred by the telltale signs of gouges made by trailer hitches and other low parts of automobiles and trucks traversing them.

Two of the speed humps on Dover Road bracket a house in which several small children live, and that may have been the condition that warranted the installation of the traffic control devices in the first place. The locations of the other humps are spread out, and one of them—the last to be installed—appears to be intended to slow down vehicles approaching a tricky curve and intersection in one direction and a busy clubhouse area in the other. In all cases, the speed humps achieve their goal of alerting speeding traffic to slow down and reminding traffic abiding by the speed limit to be cautious.

Yet, however well they achieve their primary design objectives, speed humps also have a downside. Many cars slow down more than necessary when approaching a speed hump, which means that they must accelerate back up to speed once they clear it. This not only introduces unnecessary noise (in addition to whatever noise might be generated by the passage of the car over the hump), but it also consumes additional fuel and puts out additional exhaust, thereby contributing to the problem of local air pollution and global greenhouse gases. There are critics of speed humps who claim furthermore that they distract drivers negotiating the obstacle and—contrary to their intended purpose—so endanger any children playing or adults walking nearby.[7]

Additional negative effects of speed humps include slowing down fire, ambulance, and other emergency vehicles, thereby lengthening response times that conceivably might make the difference between life and death. The presence of speed humps on one road may divert traffic to an alternate route and bring with it all the problems associated with increased traffic on a previously quiet street, which soon may also be fitted with speed humps. Then there is the question of aesthetics. The introduction of a speed hump is often accompanied by the painting of warning lines or colors on the device and the road surface leading up to it, the erection of signs alerting motorists to the upcoming obstacle, and the possible installation of more street lighting in the vicinity of the hump. Such changes to a previously quaint neighborhood add to the visual pollution of the street scene.

A science of speed humps and bumps could be devised. It might start with fieldwork to observe them in their setting and to collect data on their size and shape. Plaster casts of representative ones might be made and sent back to a central museum, where they could be catalogued, classified, and studied by resident bumpologists. Theories of speed humps and bumps might be put forth, and missing links in the evolutionary chain might be speculated upon. But all this observing and organizing might do little to change the nature of speed humps or mitigate their negative effects on the cars and drivers that bounce over them. The bumpologists might propose erecting warning signs giving the exact distance to each speed hump and the mathematical formula describing its profile, so that approaching vehicles with onboard computers could calculate how to adjust their speed to minimize the effect of the hump. This might all be theoretically possible, but engineers might prefer to redesign a particularly obnoxious speed hump or design from scratch an alternative to it.

The speed hump or bump would seem to be about as simple a designed object and as good a metaphor as can be for complicating surprises down the road. To avoid adverse effects of the bump, the road leading up to and away from it must be properly designed and built. The bump itself—its cross-sectional profile, whether derived from theoretical principles or based on experimental trials—must embody a shape that accomplishes the objective. Many an engineer will not look beyond the principal design objective, which is to slow down traffic, and so the simple consequence of the speed bump's operation may never be considered: what goes up one side of the bump must come down on the other side. After all, the design problem was not to raise and lower a car gently, but to hinder its progress. The driver would have to worry about comfort and damage. Had an engineer been asked to design a speed bump or hump that would slow down a car approaching it *and* not cause noisy operation or encourage subsequent acceleration, the engineer would certainly have explored further how to accomplish those ends. But such a design would likely follow a systems approach, where the speed hump was viewed not just as a bump in the road but as a component of an ensemble that includes also the characteristics of the cars and drivers, the nature of the neighborhood, and the expectations of the neighbors. In any case, the downsides of speed bumps suggest that unintended consequences can accompany even the most elementary and elegant of designs.

Although a lofty skyscraper might seem to be in a completely different category from a lowly bump in the road, the two share an engineering design process that can be any made thing's undoing. Making a speed bump may appear to be only a matter of mounding up hot asphalt and waiting for it to cool, but in fact there is much technology to be applied and good practice to be observed if the thing is to function properly and not deteriorate under the wheels of the first few cars that pass over it. Building the tallest building in the world presents different and greater technical challenges, some as mundane as employing cranes that can climb upward as the structure rises and mixing and pumping concrete that will not harden before it is lifted into its final resting place. But the secret to making possible a building like Burj Dubai—the tallest structure in the world long before it was even completed as a building—was keeping it relatively simple in form and process, so that repetitive construction tasks would not only make the project feasible from the point of view of its architect, the firm of Skidmore, Owings & Merrill, but also provide experience that could be built upon. William F. Baker, the lead structural engineer of Burj Dubai—which may in time reveal some unexpected speed bumps on the road to full occupancy and operation—said of its design and construction that "making something simple is never easy." The independent architect/engineer/artist Santiago Calatrava, whose bridges and buildings are considered works of art, has gone even further, saying that "to do simple things is very difficult."[8]

New York City is neither simple to build in nor easy to drive around. Its streets have few speed bumps of the conventional kind; potholes, which might be considered inverted bumps, serve to keep some of the traffic under control. But for all of New York's roughness, blare, glitter, and glare, world-class architects still prize a commission to design a new addition to the city's skyline. One such opportunity was given to the Italian architect Renzo Piano, and it resulted in what was described as "the most significant new building to be designed for the NYC skyline in decades."[9] The structure is the New York Times Building, located in a formerly seedy section of Manhattan across Eighth Avenue from the Port Authority Bus Terminal and down the block from Times Square.

This skyscraper was much anticipated by architecture critics. Paul Goldberger, writing in *The New Yorker* when the building was only a plan on paper, described it as a "lyrical design" that "appears to be lighter and more elegant than even the most pristine International Style buildings in the city—almost ephemeral, a vast structure with the delicacy of mist." The light, ephemeral, and delicate qualities of the building are largely a result of its unusual facade, a first of its kind.[10]

In order to make the building, and not incidentally its onetime owner's newspaper operation, appear transparent, it is glazed floor to ceiling with "ultra clear low iron glass" that makes the inside highly visible. But such a transparent structure could also become unbearably hot to work in when the sun is shining brightly, and so at some distance from the glass it is "draped in ceramic tubes to create a curtain wall that reflects light and changes color throughout the day." This mostly silvery-gray ceramic sunscreen, which has been described as a "filigree screen," extends up beyond the roof of the structure to add to the delicacy of which Goldberger wrote. The novelty of this feature became one of the most discussed and admired aspects of the structure.[11]

After the new home of the *Times* was completed in 2007, the false wall of horizontal ceramic rods also caught the attention of a small group of daredevils who engage in "buildering," which is the name its practitioners have given to the activity of scaling tall man-made structures. Whereas to a layperson the Times Building might appear to be sheathed in a massive set of Venetian blinds, to a builderer like the French stuntman Alain Robert it looked as if it were fitted with a ladder on which to climb the skyscraper all the way to the top. He assessed the feat to be easy, and "on a difficulty scale of 1 to 10, he rated it a 1." However, since such a stunt is, to say the least, officially frowned upon, he had to plan his attempt surreptitiously. He scouted the situation weeks ahead of time and made a quick and undetected test climb early one morning. At that time, he established that the rods were strong enough to handle his weight.[12]

Robert made his public assault on the building in broad daylight. Since the ceramic sunscreen began about twenty feet off the ground, he first had to shinny up a column from the sidewalk to an overhang, where he could mount the "urban ladder." Once he was on it, he climbed quickly so that he would be out of reach of fire department equipment, which could have been used to foil his plan. Once

beyond that point of no return, he paused to unfurl a banner promoting his environmental cause of combating global warming. Then
he continued his climb to the roof, where police were waiting to
arrest him.[13]

Later that same day, Renaldo Clarke, an amateur climber from
Brooklyn, scaled the "very wide ladder" of the same structure. He

*The New York Times Building was a much-heralded addition to the city's sky-
line, but the skyscraper's ceramic-rod sunscreen provided an inadvertent ladder
that enabled daredevils to scale the exterior of the structure with relative ease.*

had been contemplating it over the course of two years, and when Robert had "eaten off the plate of Mr. Clarke's imagination," he determined to reach skyward also. And he met the same fate, ending up briefly in the same jail cell as the Frenchman. Thus Clarke, the climber who did it to call attention to malaria, had a brief conversation with Robert, the one who did it to protest global warming. After the two successful feats, guards were stationed around the building's perimeter, and gaps in the overhang that allowed the climbers to gain access to the ceramic ladder were closed off with plywood while a more permanent solution was sought.[14]

The Manhattan district attorney brought charges of reckless endangerment, trespassing, and graffiti writing, as well as disorderly conduct against Robert, but the grand jury didn't support the misdemeanor charges and advised that only the last, the noncriminal charge of disorderly conduct, be levied. During Robert's testimony before the grand jury, he asserted that he had not put anyone at risk because he had made careful preparations for the climb. Before he mounted the structure, he distributed flyers warning people to stand back from the building. He also mentioned that he had tested the strength of the rods beforehand.[15]

While the district attorney's office deliberated over how to charge Clarke, a new grand jury was empaneled, and it voted to indict him for reckless endangerment, a charge punishable by as much as a year in prison. Speculation was that Robert's documented history of successful industrial climbs, performed worldwide in order to call attention to global warming, made him more credible and his act more pardonable to the grand jurors. According to one of Robert's Web sites, thesolutionissimple.org, "Emissions are still climbing. So am I. But the solution is simple: 1—Stop Cutting Down Trees. Plant More Trees. 2—Make Everything Energy Efficient. 3—Only Make Clean Energy." Clarke had no such focused plan; he was expected to pursue a plea deal.[16]

Shortly after Robert and Clarke had climbed for their causes, the New York City Council, frustrated that there were so few serious penalties for climbing buildings, considered a bill to criminalize the scaling of or jumping from one taller than twenty-five feet. But within five weeks of the first scalings of the Times Building, before an effective physical deterrent could be implemented, a third builderer was able to mount the accidental ladder and climb it to about the tenth floor. While resting there, the Connecticut resident David Malone used his cell phone to call a newspaper editor and was persuaded to abort his climb in exchange for a face-to-face interview. Ironically, the editor worked not for *The New York Times* but for the city's tabloid *Daily News*. Like the other climbers, Malone also had a cause to promote. He wished to bring attention to his self-published book, *Bin Laden's Plan: The Project for the New Al Qaeda Century*, on which he had worked for years but which had not attracted much attention.[17]

Within hours of the third climber incident, workers began removing ceramic rods located near the base of the building. It was expected that the lowest nine feet of the sunscreen/ladder would be permanently removed and replaced by glass panels, so that potential climbers could not easily reach the new bottom rung. When asked by reporters covering the initial climbing events, architect Piano had declined to be interviewed about his building's design, explaining that a vice president had explicitly directed him not to talk with the paper's reporters. Now, after the third scaling of his building and the subsequent modification of its facade, Piano let it be known that he was in total agreement with the solution. The striking design feature of the horizontal ceramic-rod sunscreen had proven also to be a now obvious design flaw.[18]

In the wake of the first two climbs, a vice chairman for the Times Company said that he did not recall there being, during the planning stages for the building, talk of anyone possibly scaling the fifty-two-story structure. Since work on its design was begun before the terrorist attacks that destroyed the Twin Towers of the World Trade Center in 2001, questions of security later came up and drove some design changes in the Times Building. At the same time there was a desire to create a skyscraper that did not look fortresslike, but rather was "open to the city." According to Piano, "This problem of climbers is honestly something we didn't think about."[19]

But the attractiveness of New York's tall buildings to climbers is legendary. In the 1933 movie *King Kong*, the giant ape climbed the Empire State Building, then just two years old. In the 1976 movie of the same title, the ape scaled the World Trade Center, which only a few years earlier had surpassed the Empire State as the city's tallest building. A remake of the original film, done in the wake of the 2001 collapse of the Twin Towers, again had King Kong scaling the Empire State Building. In fact, the attraction of such structures to real-life daredevils had been apparent as early as 1974, when the Frenchman Philippe Petit walked back and forth on a cable strung between the tops of the Twin Towers. The appeal to human builders had been evident at least as early as 1977, when a mountain climber used specially designed clamping devices to ascend the South Tower of the World Trade Center. After that, architects had tried to avoid details that made it easy to climb tall buildings—except in the case of the Times Building. Even the base of the new Freedom Tower, the centerpiece of the World Trade Center being rebuilt around Ground Zero, was being designed without ledges that might provide a leg-up to climbers.[20]

Given the history of daredevils and tall buildings in New York, it should not have taken much imagination on the part of an architect—or the client reviewing the design—to see how a curtain wall of horizontal ceramic rods could double as a ladder. Had this possibility been considered and acknowledged, then the design (or redesign) problem would have become one of changing the arrangement of the rods to obviate their being used as rungs. An easy change would have been to install the rods vertically, but this may not have pleased the architect's aesthetic sensibility or as effectively served the sunscreening function. Another solution might have been to enlarge the diameter of the rods so that a human hand could not be wrapped around them. This also might have offended the architect's aesthetic goal and frustrated his functional one. Perhaps the rods could have been placed closer together, so close that a foot could not be inserted between them, thus precluding climbing. If the architect did not like this, perhaps because it cut down on the amount of light that was let through the screen, then other geometric variations and compromises could have been explored—until the right combination of appearance, function, and prevention was achieved.

.   .   .

If such simple things as speed bumps and a building facade can reveal unanticipated negative features only after they are installed and in use, and if they can surprise their designers and owners with unintended downsides, then how many more undesirable consequences might be latent in more complex devices and systems, including those designed with the best of renewable energy and environmental intentions? And how long might it take for them to reveal themselves? These are especially important questions for engineers to ask as they devise schemes intended to reverse the effects of such planet-wide problems as greenhouse gas buildup and meet other long-term global challenges, for sometimes unforeseen speed bumps can arise imperceptibly slowly in an apparently necessary and benign technological system—until the problems they present appear almost insurmountable.

At the end of the nineteenth century, the horse had long been the standard transportation and work machine (and the output of an automobile internal combustion engine is still measured in horsepower). In the United States the number of horses was about thirty million, or one for every four people. It was more or less taken for granted that the indispensable horses required a considerable investment in infrastructure and support staff—including barns and grooms and farriers and waste removal systems—and it has been estimated that the production of horse feed required about a third of the nation's farmland. Annually, the dung from each horse provided tons of manure, which was good in farming country but inconvenient, to say the least, in urban areas. Horse pollution had made daily life in the Victorian city a dirty, smelly, unhealthy, and generally unpleasant experience, and so replacing the animal became an increasingly desirable objective.[21]

There were a variety of alternatives from which to choose. By the end of the nineteenth century, bicycles had evolved into highly mobile clean machines, and an electric vehicle held the world speed record of sixty-one miles per hour. By the turn of the century almost as many electric-driven cars (1,575) as steam-driven ones (1,684) would be manufactured annually in the United States. Combined, they outnumbered gasoline-engined cars by more than three to one. The electric vehicle had the clear advantage of quietness over its then unmuffled competitors, and it did not need to be hand-cranked to be started. However, perhaps at least in part for sentimental reasons, Henry Ford preferred the gas-fueled motorcar. He once presented his wife with a "design for an internal combustion engine, drawn on the back of a piece of music," and successfully tested one about two years later. At the time he was working at the Edison Illuminating Company, and he continued to dream of making a complete automobile. Ford accomplished his dream in 1896. When Thomas Edison heard about it, he encouraged the young mechanic. But, ironically, Edison "was adamant that Ford not waste his time trying to make a car run viably on electricity." Edison himself had come up with a nickel-iron battery, but had found its cost to be prohibitive.[22]

Ford took Edison's advice to heart. After the introduction of the Model K in 1906 and the Model T in 1909, the internal combustion engine became the power source of choice. By 1912, there were 900,000 gasoline-powered vehicles in America, outnumbering the 30,000 electrics thirty to one. At about the same time, the self-starter and silencer (muffler) were introduced, thus making the internal combustion engine more user-friendly and less objectionable. Even if extra canisters of fuel had to be carried on a drive of any distance, it was so much faster to add gasoline to a car than to recharge heavy batteries that did not give it as much of a range as a tank of gas. (Different energy sources are now compared by means of a measure known as energy density, which is the ratio of energy to mass. A typical conventional lead-acid battery might have an energy density of about 35 watt-hours per kilogram compared to gasoline's 2,000. Although more exotic types of batteries have higher energy densities than lead-acid, those densities remain an order of magnitude smaller than that of gasoline, and the lighter batteries are more expensive to manufacture.) The use of gasoline-powered vehicles in World War I conditioned a lot of young veterans to favor the internal combustion engine. The last new model of an electric car to be built in America during that era was introduced in 1921—at a price four times that of a Model T.[23]

The electric vehicle essentially went into forced hibernation for many decades, until environmental awareness and energy crises reawakened interest in a nonpolluting alternative to the internal combustion engine. According to environment-conscious critics, the carbon emissions from gas-guzzling internal combustion engines were contributing to a greenhouse effect of global proportions. What had seemed to be a horse-dependent society's salvation from aesthetic, health, economic, and environmental points of view in the early twentieth century was seen as a threat to the very planet Earth as the end of the century approached. No one wanted to go back to horse-and-buggy days, but powered vehicles of all kinds had multiplied well beyond the horse population of a century before, and the infrastructure of roads, parking lots, gasoline stations, service stations, and oil refineries—and an international petroleum cartel—made the horse era seem truly quaint. Perhaps the problems being experienced a century later could have been foreseen when the automobile was in its infancy, but even if they could have been, it is unlikely that the conversion from horse to horseless would have been curtailed.

One of the most eagerly anticipated new transportation technologies of the opening years of the twenty-first century was known by its code name, Ginger, while it was under development and simply as It in the months of tantalizing hype preceding its unveiling in 2001. Among the early endorsers of It was Apple Computer's Steve Jobs, who was reported to have said that "the device might change urban life and could be as significant as the development of the personal computer." Ginger turned out to be what is now known as the Segway Personal Transporter, the brainchild of inventor Dean Kamen. Looking somewhat like an old push lawn mower, but with a platform in place of the blade assembly, the Segway supports a single standing rider, who holds on to the high handlebars and controls the battery-powered machine by leaning slightly forward on them to go, sideways to turn, and backward to slow down and to stop. In the stopped position, the Segway platform maintains its horizontal position and the handlebar post its vertical by means of individual-wheel motors controlled by computer electronics wired to gyrocompasses.[24]

Kamen's vision for Ginger—named after Fred Astaire's dancing partner—was that it did not belong in the street: "We are an *electric pedestrian*. It's *not* a vehicle. It's an alternative to walking. We *can't* let them call us a scooter." Small, electric-powered scooters (with wheels in tandem) had recently experienced a jump in popularity, but they were also viewed in some communities as a hazard on sidewalks. What the Segway was called and how it was classified were important, because different jurisdictions had a maze of different laws prohibiting powered vehicles from riding on the sidewalk, as well as regulations requiring them to have such accessories as brakes, horns, turn signals, and reflectors, not to mention license plates and insurance. Anticipating all the permutations and combinations of the potential speed bumps on the road to wide adoption of the Segway was mind-boggling. Lobbyists were hired to influence local and state legislation regarding powered vehicles and sidewalks.[25]

As of 2004, the company claimed that scooters were legal on sidewalks in forty-one states, and in these states Segway did not mind its machine being called a scooter. But larger cities with crowded sidewalks, like San Francisco and New York, did not think pedestrians and Segways mixed well, at least when private citizens rode them. Although production facilities were supposedly capable of making forty thousand personal transporters per month, early sales were well below that rate. The device was said to be "incorrectly positioned" in the marketplace. According to the author of a textbook on product design, "By putting Segways on sidewalks, the company is saying their transporter is just like walking, but better" (it can reach a speed of twelve and a half miles per hour), ignoring the fact that walking is healthier. However, promoting riding a Segway as a substitute for walking implied that it was for relatively short trips, which did not represent a significant niche market.[26]

Another bump in the road to market acceptance came when the machine that television news anchorman Dan Rather said "just might be the most revolutionary leap since the automobile" could not always make it over a physical bump. If the batteries were low when the rider called for a sudden burst of power, such as would be needed to speed up or clear an obstacle, the machine could simply stop and topple over. Naturally, the rider was in danger of falling over with it and suffering injury. Although the Segway displayed multiple warning signals when the batteries became low, which occurred after about twenty-four miles, users ignored the warnings. The unanticipated bug led to an embarrassing recall less than a year after the machine debuted. Software upgrades were made to provide earlier warning signals and shut the transporter down before adverse behavior could result.[27]

By the summer of 2006, almost four years after it went on the market, Segway sales reportedly stood at tens of thousands worldwide, with 40 percent of the machines going to police departments, tour guide companies, and other commercial firms. Not only did the transporters allow officers on the beat to cover more ground faster, but also the law enforcers looked taller and achieved an "imposing stature" when standing on the machine's platform, which raised them eight inches off the ground, whether they were riding in the gutter or on the curb. Where Segways ridden by civilians are banned from sidewalks, many policemen turn a blind eye to the practice. One tour director in New York City, who acquired his first Segway in 2003 and uses the transporters to guide tourists along the city's sidewalks, reported that officers of the law had "stopped him only to ask for a ride." Still, as of the summer of 2008, it was estimated that only about thirty private-citizen New Yorkers owned Segways. The general population of the city did not see the machines as replacements for walking on its crowded sidewalks.[28]

In fact, most innovations are seen clearly to be a new solution to an old problem. When a new technology is evolving, it typically starts small compared with whatever it is to replace, and can be largely ignored until it becomes an obvious threat to the status quo. By that time, it has its own momentum and support to push against any organized opposition by means of legislation or regulation. It is no easy thing to anticipate the long-term downside in pollution and congestion of such a complex system as the fossil-fueled automobile and the associated network of roads and service stations and other infrastructure, which evolve not only along with but also independent of the main innovation. But for simpler designs, it is reasonable to expect an engineer to anticipate what could go wrong. As much as possible, an inventor or designer must be of two minds: thinking about the immediate problem at hand, while at the same time looking into the future for unexpected, undesirable consequences. And sometimes that future is not too far away.

Only decades after e-mail was introduced, a front-page headline in *The New York Times* identified the electronic messaging system as a

"self-made beast" with which high-technology companies were trying to cope. Specifically, as the article explained, firms like Microsoft,
Intel, Google, and IBM, which had responded to a survey of businesses that were considered information intensive, found that large
chunks of time were being wasted because of distracting e-mail messages. More than a quarter of a typical information worker's day was
consumed by interruptions caused by unnecessary e-mail messages
and cell phone calls, while only about an eighth was available for
thoughtful and reflective work. One analyst involved in the survey
spoke of the companies facing "a monster of their own creation" and
cited a maxim of Silicon Valley saying that companies should "eat
their own dog food," meaning that they should use what they put out
for consumers. The problem was that they were, and that they were
"eating too much."[29]

It did not take managers or consultants to identify the problem;
many among the rank and file had long since become acutely aware
of it. And in true Silicon Valley scientific-engineering style, they had
devised appropriate jargon and slang to characterize various aspects
of information overload and its undesirable effects. The condition of
letting e-mail messages pile up to the point that the only option is to
delete the lot of them and begin anew came to be known as "e-mail
bankruptcy." Then there is the phenomenon of unconsciously holding one's breath upon discovering an unmanageable number of new
messages. This came to be known as "e-mail apnea."[30]

What was bothersome was not so much the time consumed in
reading a just-arrived e-mail message as the time it took to get back
to what a worker was doing before the interruption. As with the
speed bump, it is not so much the slowing down for the bump itself
as its enduring effects after the bump has been passed—whether in
the jarring letdown or the need to get back up to speed. To deal with
the Frankenstein known as e-mail, affected companies formed the
nonprofit Information Overload Research Group to discuss solutions. Even before the group met, however, some approaches were
being tried. An IBM engineer named Michael Davidson devised an
"E-mail Addict" feature whereby a user could click on a "Take a
Break" link that caused the screen to go gray and display the message, "Take a walk, get some real work done, or have a snack. We'll

be back in 15 minutes!" There seemed to be no acknowledgment that taking a walk or having a snack might be more disruptive than receiving another e-mail message. Nor was it explained how one gets some real work done with a gray computer screen.[31]

Meanwhile, the BlackBerry blurred the line between work and leisure. This issue arose at ABC News, where there was a long-standing agreement between the company and news writers that they would not be paid time and a half for overtime just for checking a company-issued BlackBerry. When ABC asked some newly hired writers to sign a formal waiver to this effect, there was concern that this could mean that employees might not be paid for using their BlackBerrys outside working hours to do more than just check e-mail. Until the speed bump was cleared away, ABC relieved the new employees of their BlackBerrys.[32]

In a much-discussed article in *The Atlantic*, Nicholas Carr, a commentator on technology, business, and culture, argued that habitual use of the Internet and Web was changing the way people think and read—and changing it for the worse. He took as one of his prime examples Google, which he described as "a fundamentally scientific enterprise" that wished to "solve problems that have never been solved before." Carr also asserted that the problem that the founders of Google wished to solve, that of creating artificial intelligence, was the "hardest problem out there." Of course, such an ambitious endeavor would not be a scientific but an engineering enterprise.[33]

Carr's essay was criticized by *New York Times* writer and blogger Damon Darlin, who speculated that negative views of new technologies could result from the viewer's training or profession. He cited the futurist Paul Saffo, who has divided the world of technology into two camps: engineers and natural scientists. According to Saffo, engineers by nature have an optimistic worldview; they believe that "every problem can be solved if you have the right tools and you pose the correct questions." Scientists, to whom he attributes a pessimistic view, "see the natural order of the world in terms of entropy, decline and death."[34]

The test for telling the difference between an optimist and a pessimist is often taken to be whether an individual sees a glass containing equal parts of air and water as half full or half empty. There is a joke that says that the engineer is neither: an engineer sees the glass as improperly designed, for it is too large for its evident purpose. But jokes are at the same time half-truths and half-untruths. Engineers are indeed optimists, in the sense that they believe that they can solve any problem, given enough resources. And some scientists, at least, are indeed pessimists, in the sense that they see the end of the world even in its beginning—the Big Bang.

Scientists also warn us of the entropic disasters associated with climate change, asteroid strikes, and the like, but warnings are not solutions—nor are they necessarily a death knell. It will be the optimistic engineers who hear the warnings not as doomsday scenarios but as calls to tackle significant problems. They usually will need to do more than just ask the correct questions and have the right tools, however. They will also need lots of time and money in the case of mounting such out-of-this-world missions as covering an incoming asteroid with mirrors to catch the sun's deflecting rays. And it would also take a good deal of reflective material that could be folded into rockets and then deployed on the surface of the asteroid. Engineers recognize that solutions to far-reaching problems will not necessarily be easy, nor can they be expected to be achieved without setbacks. Just as the world's largest and most complex machine, the Large Hadron Collider, exhibited embarrassing glitches upon being started up for the first time, so will planet-saving schemes likely experience setbacks upon being implemented. This will, of course, mean that more time, money, and materials might have to be expended if our planet is not to perish.

If a relatively small thing like a speed bump cannot easily be designed without accompanying adverse consequences, how much more difficult must it be to design, develop, and debug more complicated machines and systems like gigantic particle accelerators, not to mention finding worldwide and even extraterrestrial solutions to global problems. It will take a lot of concerted effort, and a commensurate amount of patience—on the part of scientists, engineers, law- and policy makers, and the public alike—in dealing with obstacles, setbacks, and bumps in the road to true energy independence and any other grand goal that is so much more easily stated than achieved.

Even if the bumps in the road toward realization or the downsides of a new piece of technology can be foreseen, it does not necessarily mean that a concept is a bad idea. It is the essence of engineering to anticipate potential problems with a design and to redesign accordingly. Unfortunately, sometimes the warning signs are not heeded, and potential problems are dismissed with statements like "That will never happen if the technology is used correctly." Such wishful thinking does not belong in the mind of an engineer or architect. The ironic nature of poor design is that it accentuates the positive but often fails to eliminate the negative—or to acknowledge it, if it is even recognized—beforehand. The best design acknowledges the existence of speed bumps, watches out for them, anticipates them, negotiates them, analyzes them, and designs ways around them.

The horseless carriage evolved in a very diffuse manner over a period of decades, and the focus of the pioneer designers and manufacturers was on getting their machines to work better than those of the competition, whether measured by performance, price, or some less quantitative metrics like aesthetics, style, and status. Speed bumps were taken in stride. Many of the early automobile manufacturers were small enterprises, working out of carriage houses that would in time be replaced by the now proverbial garages. At least initially, workforce and output were small. In older and larger industries there was a more managed process of developing a new product. The electrical devices manufacturing industry was growing by leaps and bounds and consolidating by the time automobiles began to come one by one out of workshop-garages. And it was in the more mature industries that organized research and development with the goal of conceiving, designing, and manufacturing new products to maintain a competitive edge began to be institutionalized. This, at least in principle, would bring scientists and engineers to work together with a single goal as perhaps never before. It was not always a happy or fruitful marriage. And carefully considered new products were not necessarily without latent flaws and speed bumps of their own.

7

Research and Development

Generally speaking, research and development is just another name for science and engineering, but jointly directed to some corporate, institutional, or national goal, and typically—in industry at least—involving new or improved products or processes, ideally with no unintended consequences. The concept of the research and development laboratory dates from at least the latter part of the nineteenth century in America, and it may be said to have gained prominence with the very visible efforts of Thomas Edison. In 1876 he set up at Menlo Park, New Jersey, what is considered the first industrial research laboratory, whose achievements he was not shy about promoting. The rural setting put the lab far from any factory, but communication and transportation links connected it to cities and their financial, business, and media resources. In 1887 Edison relocated operations to West Orange, New Jersey, and designated the works his "invention factory." He claimed that the complex, which included laboratories, shops, and a library, contained everything that was needed to invent "useful things every man, woman, and child in the world wants . . . at a price they could afford to pay." But whether the Edison Electric Light Company could afford or wanted to fund the escalating costs of research based on the direct-current principle was a question that did occur to the firm's directors. Money and results, or the lack thereof, are constant themes where research is involved, whether the task is to develop low-cost LEDs today or dependable incandescent bulbs more than a century ago.[1]

Edison Light was absorbed into Edison General Electric, which by merger in turn became part of the General Electric Company; its competitor was the Westinghouse Electric Company, which pursued the alternating-current route to success. From its founding in 1886, Westinghouse had engaged in research to develop new products and create new knowledge about materials and electrical machines. The work in what would become the laboratories of GE and Westinghouse has been described as "engineering research," which favored product development and materials testing without paying much attention to what was called "fundamental research" or "fundamental science," more commonly known in government laboratories as "basic research" or "basic science." The companies understood that there was no need to engage in pure scientific research to ensure engineering achievements.[2]

Whereas the electric companies were, at least initially, more engineering-driven, the pioneering corporate laboratories founded in the first decades of the twentieth century, along with those of the electric companies, increasingly engaged in fundamental research that complemented but did not replace the testing of materials and the development of products. Among such laboratories were those associated with companies like American Telephone and Telegraph, DuPont, and Eastman Kodak, as well as the Aluminum Company of America and Corning Glass Works, all of which established research labs to achieve one or more specific objectives. According to historians of technology Ronald Kline and Thomas Lassman, these objectives were to "mitigate market competition, avoid antitrust suits, internalize the process of technological innovation, and diversify their product lines." Just as different companies have different corporate cultures, so did the research and development programs that they set up, each with different organizational strategies to reach their goals. Sometimes they followed the strategy of an indispensable eccentric like Charles P. Steinmetz, who was notorious for smoking cigars where there were NO SMOKING signs and appearing in work clothes when the occasion called for more formal attire.[3]

Steinmetz, whose name would become synonymous with "engineer," began working for General Electric in Schenectady, New York, in 1892, and soon rose to the custom-made position of chief consulting engineer. In that capacity, he was expected to troubleshoot major technical problems encountered within the company and advise top management on technical issues and opportunities. In the late 1890s he began advocating for a research laboratory, something he felt that GE needed if it wished to continue to be competitive and to grow into new areas. When his recommendation was finally accepted, in 1900, an electrochemical research laboratory was set up to develop new types of lamps, thereby helping GE withstand the growing competition from the likes of Westinghouse. Today, General Electric claims that the first corporate research laboratory sponsored by U.S. industry was established "in the carriage barn in Steinmetz's backyard."[4]

Steinmetz became legendary in his own lifetime, and countless photographs were made of him among his laboratory apparatus, as well as of his visiting and being visited by the likes of Edison, Guglielmo Marconi, and the popular screen actor Douglas Fairbanks. General Electric was so eager to exploit photos recording such meetings that its publicity department became overzealous in its attempts to secure shots with even greater impact. In 1921, Steinmetz and Einstein were standing beside each other among a group attending a demonstration sponsored by the Radio Corporation of America of transoceanic telegraph transmission at New Brunswick, New Jersey. Long before digital photography, the group shot taken was cropped and doctored by the GE publicity department to show only the scientist and the engineer, as if they were meeting one-on-one. The widely distributed photo can still be found reproduced without a hint of its true origins. As I have described elsewhere:

> Unlike many of the other photos of Steinmetz, which appear to show him posed or caught in positions of mastery and authority that deemphasized his congenital hunchback and stooped posture, the photo with Einstein seems to emphasize both. The father of Relativity stands tall, formal, and self-assured beside the short and stooping, hand-in-pocket and smoking, rumpled and scowling figure who looks to be, ambivalently, both leaning toward the famous scientist and pushing him back with an elbow. The incongruity of the pair is accentuated by the juxtaposition of the hatted Einstein's dark overcoat and the bareheaded

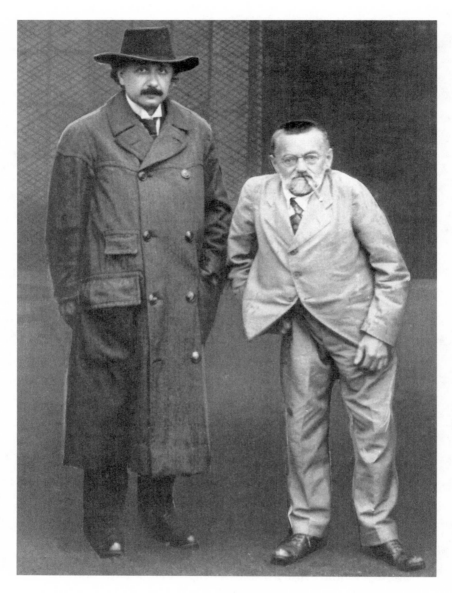

A widely distributed photograph of a purported meeting between Albert Einstein and Charles Steinmetz provides a striking contrast between the public images of scientist and engineer. In fact, the figures of the two men were cropped from a group photo to suggest for publicity purposes that they met one-on-one.

Steinmetz's light suit, as if the two were not even standing in the same weather.[5]

The doctored photo of Einstein and Steinmetz may also be seen as a metaphor for the scientist and the engineer, for science and engineering, and for dysfunctional research and development, at least to an outsider.

Research and development is commonly abbreviated R&D, and the conjunctive ampersand can be interpreted to be symbolic of how curious and cryptic the connection between research and development—that is, between science and engineering—can be. Indeed, it is the nature of the *&* in R&D that can make the difference between the success or failure of the enterprise.

Essentially a ligature for *et*, which is Latin for *and*, the ampersand is among the greatest curiosities of typography. In fact, at one time *&* was included as a part of the English alphabet, coming after the letter *z*, and children would conclude their recital of their ABCs with "... *X, Y, Z, & per se, and.*" The seemingly unpronounceable finial *&* was in fact pronounced "and." Over time, "and per se, and," which was often slurred in speech, became corrupted into the word *ampersand*.

Typeface designers seem to have been fascinated with the ampersand, sometimes making its Latin origins all but explicit, as demonstrated quite strongly by the & of the face Trebuchet MS and to a lesser degree by the & of French Script MT. Ampersands in most modern typefaces do not reveal their etymological roots so clearly, but once one knows what to look for, it is possible to see a stylized *Et* in virtually all of them, as is the case with the familiar & of the sans serif typeface Arial and the & of Times New Roman. Whatever their family ties, many of the most common ampersands resemble an illustration one might expect to find in a how-to book on knot making— or someone tying a shoelace. The knot- or bow-tying image is apt, for the use of the ampersand in R&D emphasizes that in the best of research and development, the science and the engineering are fastened together in a way that ties them into a single effort.[6]

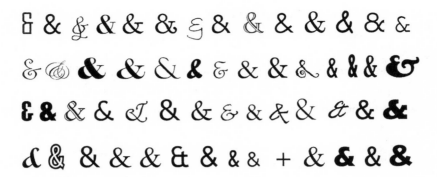

This assemblage of ampersands from a wide variety of typefaces is suggestive of the many ways in which research and development can be connected and interact in the collaborative scientific and engineering endeavor known as R&D.

I have always admired the cleverness of the aptly named RAND Corporation. It was established within the Douglas Aircraft Company in 1946 as Project RAND, which was funded by the U.S. Army Air Forces to engage in the study of future weapons development. When it became evident that there might be a conflict of interest if the project recommended new weapons for which Douglas would be the likely manufacturer, the nonprofit corporation was spun off from the aircraft company. After its incorporation in 1948, this "seedbed of systems analysis" focused on a wide range of issues related to national security. But whereas the word *RAND* when spelled out sounds virtually indistinguishable from *R and D*, it does not through its spelling or sound carry the same symbolic uniting force that the ampersand in *R&D* does.[7]

Just as there are seemingly as many different ampersand designs as there are typefaces, so there are about as many different industrial and government styles of research and development as there are corporations and government agencies. According to Kline and Lassman's study of "competing research traditions in American industry," there was a plethora of different models in the early twentieth century. Thus, the central laboratory of GE was distinct from the company's engineering groups, but the research unit at AT&T, which eventually became the Bell Telephone Laboratories, was within the engineering department. Different models prevailed in the chemical and metals industries. DuPont's first scientific research laboratory, which was established in the early years of the twentieth century, was charged with developing new explosives and diversifying product lines. In 1908, Corning established a small laboratory for testing materials, solving production problems, and developing new products; eight years later, the lab was expanded with the expectation that it would do more fundamental research, a goal that was not to be met until after World War II. In the meantime, at Eastman Kodak, where an experimental and testing laboratory was already located close to the manufacturing facilities, an independent large-scale fundamental research laboratory had been established in 1912. Alcoa's diffuse research efforts were brought together in 1919 with the establishment of a technical department modeled after the research lab at GE.[8]

Wherever a laboratory was located, among the continuing topics of debate was the role and value of fundamental research. Even when there was a commitment to it, the question remained about how to organize a research and development laboratory. Was the *R* or the *D* more important in *R&D*? And how was the conjunction—the ampersand—to be handled? Naturally, different companies reached different conclusions. At both Westinghouse and AT&T, for example, scientists were supervised by an engineering department, but such traditional engineering research activities as the testing of materials exerted a much greater influence on what research was done and how it was organized at Westinghouse than at AT&T. In general, according to Kline and Lassman, "the interactions between a long-standing engineering culture and a culture of fundamental science imposed from the top down created a contested research culture." This can be best understood in the context of extended examples.[9]

One of the historically significant figures in R&D at Westinghouse was the electrical engineer Charles E. Skinner, who in 1906 was made head of the company's research division. He referred to the work of his division as "commercial research," which to him included testing materials, developing new processes for treating materials, improving designs, developing new apparatus, troubleshooting failures, and looking into unsolicited "crank schemes, scientific fakes, etc." Skinner did not consider fundamental research to be an activity of his division, explaining that those on his staff were not "privileged to study science for the pure love of science, but must become part of the great industrial machine which takes scientific facts and phenomena and weaves them into a fabric of machines and devices." He saw no need to do fresh science to develop new products. However, with the advent of a world war, it became evident that there was some need for basic scientific research, and a new laboratory at a physically separate location was established to house the "theoretical" lab, as opposed to the "practical" labs located at the factory. Skinner believed that keeping the two locations under his own management would cancel any disadvantages that might have resulted from the space between the R and the D groups.[10]

Skinner was obviously optimistic about how theoretical results would be effectively translated to the practical. It would seem that to him the & in R&D suggested a relay racer carrying a baton. As in a relay race, where there has to be good coordination between runners and the pass-off has to take place within a designated lane, so in an R&D context the R and the D have to be coordinated and have to accomplish their exchange within a certain window of opportunity. One thing Skinner did realize is that quality of research mattered, and he tried to recruit top scientists to direct the new theoretical laboratory. His first choice was Robert Millikan, who would win a Nobel Prize for measuring the charge of an electron. The offer was declined, presumably because Millikan, then at the University of Chicago, wished to remain in an academic position. (Soon after Skinner's overture, Millikan moved from Chicago to Pasadena, California, where he headed up a new physics laboratory at Throop Polytechnic Institute, which was shortly to be renamed the California Institute of Technology.)[11]

Skinner next looked to a scientist already working in an industrial setting: Perley G. Nutting, a physicist who specialized in optics. Nutting had worked at the National Bureau of Standards before moving to the Eastman Kodak Research Laboratory, where he headed its physics division at the time he was contacted by Skinner. Nutting accepted Skinner's offer and moved to Westinghouse in 1916, bringing with him the model with which he was familiar: "About one-third of the time at the Kodak laboratory was spent on fundamental research, one-sixth on manufacturing problems, and one-half on developing new products." A central tenet was that "teamwork, not individual genius" was key. There was no distinguished scientist working on a particular problem with a group of researchers under him, as had become the case at the General Electric Research Laboratory. Rather, research teams were made up of scientists from different laboratory departments, and these interdisciplinary teams met weekly with production engineers and manufacturing staff, thus compressing the spaces out of R & D.[12]

Not surprisingly, over time tensions developed between Skinner and Nutting, the former pressing for commercially focused work instead of the fundamental research favored by the latter. The two could not even agree on what to call the different types of research being carried out, with Skinner using "theoretical" and "practical," and Nutting preferring "scientific" and "technical." After serving for four years as director of the Westinghouse Research Laboratory, and during that time becoming a nationally prominent spokesperson for wartime industrial research, Nutting left the company. That he did not move into another full-time position suggests that his departure was sudden and forced. For a while he worked on lighting equipment as an independent consultant to the movie industry; in 1925 he began to work in geophysics at the U.S. Geological Survey, where he spent the rest of his career.[13]

Corning Glass Works' early research problem was to tackle the company's "fundamental mystery," which was to sort out "what elements added to a basic glass mixture would produce which properties?" This was taken on by Amory Houghton Jr., son of the founder of the works and its "first researcher." The next generation of Houghtons inherited a "scientific approach to glassmaking," but they were urged by a consultant to go further by emulating the research and development efforts of two of their greatest customers, General Electric and Westinghouse, to whom they provided the glass envelopes for lightbulbs, and set up a research laboratory. This would provide the company with broader knowledge of relevant materials and processes than their competitors possessed, and so would be a long-term asset. The consultant, Arthur L. Day, head of geophysical research at the Carnegie

Institution of Washington—then the leading private research institution in the United States—concerned himself with "naturally
occurring phenomena like the melting and cooling of minerals in volcanoes." Although volcanism might seem to be remote from commercial glassmaking, Day's scientific knowledge, experience, and wise
counsel were invaluable to Corning.[14]

The brothers Alanson and Arthur Houghton were enthusiastic
about staffing a corporate laboratory with specialist scientists, but
their father, Amory, was not in favor of spending money that way. He
and his own father had kept glass formulas as family secrets, and
inviting outside scientists into the corporate fold would threaten the
continuation of that practice. The younger generation prevailed,
however, and in 1908 the brothers Houghton hired the silicate
chemist Eugene C. Sullivan to set up Corning's first open-ended
research laboratory. Since graduate-level science education was not
yet well established in the United States, Sullivan, who had attended
the University of Michigan, had gone on to further study in Leipzig,
Germany, where the world's leaders in physical chemistry were concentrated. He earned his doctorate there, and was well connected
within the field. Rather than fear the loss of family control, the
younger Houghtons treated scientists Sullivan and Day like family
members. Both would eventually become members of Corning's
board of directors.[15]

But first research director Sullivan had to set up a new laboratory.
He soon found that glass chemistry was "an unplowed field: there
was no literature and few tools or methods. New laboratory methods
and new tests had to be developed, and before much headway could
be made, measurements—of strength, clarity, and other properties—
had to be devised to measure the lab's progress." He overcame these
engineering deficiencies, and his laboratory was soon defining the
new field of "glass technology." Years later, he reflected on what
the nature of the scientists and engineers involved in a successful
endeavor should be:

> The industrial researcher . . . should above all be versatile: able
> to perform three overlapping and inseparable functions: quality
> control, improvement of manufacturing methods, and the devel-

opment of new products. To do this he needed to be broadly educated in science but also required intimate exposure to the problems that arose in the industrial heart of the business. In contrast with engineers, who built on the basis of experience, the industrial researcher needed to concern himself with fundamental principles and to conduct experiments to arrive at new facts.[16]

The laboratory established by the chemist Sullivan, augmented after a while with physicists—and all interacting with customers—was responsible for some outstanding achievements, including the invention of Pyrex cookware made from borosilicate glass, a high-quality material developed in Germany in the late nineteenth century and reverse-engineered at Corning in the early twentieth. Increasingly, Sullivan came to adopt the technical strategy embodied in the slogan "special glasses developed for specific purposes," which was to say, "glass as an engineered material." (A complete science of glass remains a puzzle to this day. According to the Nobel laureate physicist Philip W. Anderson, "The deepest and most interesting unsolved problem in solid state theory is probably the theory of the nature of glass." But this scientific ignorance has not, of course, prevented Corning from continuing to advance the engineering of the substance.)[17]

After World War I—the time when numerous companies were starting up corporate research laboratories—Corning found itself in the position of having an established but limited research lab and with the enviable problem of being able to sell more glass and glassware than its factory could produce. The Houghtons, who had tried unsuccessfully for years to lure Day to full-time employment at Corning, finally captured him in 1919, but only temporarily. He came with the expectation that he would expand the laboratory and "make a serious and sustained contribution both to knowledge about glass and to the strategic direction of the business," following the Kodak model based on "researcher freedom and the free flow of information combined with discipline and an awareness of strategic company goals." However, he soon found himself heading up not only the research but also the manufacturing side of the business and so stayed only two years.[18]

Shortly after Day left Corning to return to Washington, the company's laboratory committee discovered that only 5 percent of lab time was being devoted to research. The rest was consumed with demands from the sales department to understand customer complaints and from the manufacturing department to test and troubleshoot. This situation may have concerned the committee, but it did not necessarily hamper creativity. A Corning chemist with more than forty patents to his name recalled that the members of the scientific staff "were all pretty much glass technologists together." Furthermore, he said:

> We were given absolute control of what we worked on. We were just there and spent the day doing our darnedest to dig up something new and no one directed us to what we should be working on and put a time limit or anything of that nature. So as these problems came up, and mysteries occurred we were all extremely anxious to find out why these things happened, unusual things, and what good use could we put them to . . . there were so many obvious things to be working on and so few people to do them.[19]

The situations at the Westinghouse and Corning labs represented only two of many tentative alliances between engineering and science that played out before World War II. The players and the local issues varied, but the underlying tensions between scientists and engineers, between science and engineering, were pervasive. These tensions continue to this day in contexts large and small. Even when it comes to attacking problems of global import, unconnected to corporate objectives involving new products and profit, the disparate cultures of scientists and engineers can interfere with real progress. It is when scientists and engineers put their professional differences and egos aside and cooperate in doing research and development leading to practical engineering work that real and current problems—whether they be corporate, national, or global—can best be tackled and solved.[20]

8

Development and Research

World War II has been called the "scientists' war." From the summer of 1939, when Einstein wrote his letter to President Franklin Roosevelt, bringing to his attention the expectation that "uranium may be turned into a new and important source of energy in the immediate future" and suggesting that the president might want to "have some permanent contact maintained between the administration and the group of physicists working on chain reactions in America," it was scientists who would be the most influential players in the Manhattan Project. The success of the project in achieving its ends—fabricating the bombs that brought a conclusion to the war—would put scientists center stage in influencing postwar research policy.[1]

The Manhattan Project was in fact a supremely ambitious life-or-death research and development program charged with coming up with a new product before the competition—the enemy—did. In this sense it rivaled and eventually showed up the best of industrial R&D programs. Given that the Manhattan Project would have been considered a failure had it not resulted in the successful detonation of an actual atomic bomb, engineering was at least as important to it as was science. But the dominant positions and personalities of the scientists and the effective subsumption of "engineering" within "science" assured that the latter would surpass the former in visibility, influence, and research spoils after the war. There were some scientists who lamented what their work had wrought, but few rejected its funding fallout. Ironically, the one individual who most influenced the future course of scientific research in America was an engineer.

Vannevar Bush, who said of himself, "I'm no scientist, I'm an engineer," asserted that the war was not the scientists' alone. Indeed, he declared that "it had been a war in which all have had a part." Nevertheless, he did single out the scientists, so many of whom had put aside their professional competitiveness to engage in weapons and other war-related work, for continued attention. With the war over, Bush asked, "What are the scientists to do now?" He felt that the question was easy for biologists and medical scientists to answer, for their wartime work was not so much a detour as a continuation along their usual research path. It was different for the physicists, however, for they had been "thrown violently off stride" when they "left academic pursuits for the making of strange destructive gadgets." They had served their country and the Allied cause well as "part of a great team," and they had "felt within themselves the stir of achievement." But, Bush asked, where would they find in peacetime "objectives worthy of their best?"[2]

Vannevar Bush himself did not have to redirect his energies. He was born in Massachusetts in 1890 and, after a sickly childhood, studied engineering at Tufts College, from which he graduated in 1913. He worked for General Electric for a while, but soon went back to school and received his doctorate jointly from Harvard and MIT in 1916. As a member of the electrical engineering faculty at MIT, he developed, among other devices, a differential analyzer—a forerunner of the analog computer—to solve equations characterizing electrical power circuits. Bush advanced to dean of engineering and then vice president before leaving MIT in 1938 to become president of the Carnegie Institution of Washington, one of the largest funders of pre–World War II research. In the nation's capital, Bush also assumed his first chairmanship of a federal agency—the National Advisory Committee for Aeronautics. The NACA reported directly to the president, had its own budget, and issued research contracts to academic and research laboratories.[3]

Bush's involvement in the NACA brought him into increasing contact with defense matters, and despite being a civilian he was to become chairman of the Army and Navy's Joint Committee on New Weapons and Equipment. Bush and his confidants, who included the presidents of Harvard and MIT—the chemist James B. Conant and physicist Karl Compton—knew from their experiences during World War I that the military structure was not conducive to efficient development of new weapons, including a possible atomic bomb. The National Research Council of the National Academy of Sciences might have seemed a logical coordinator of such efforts, but Bush believed its apolitical tradition and lack of an independent budget would have made it ineffective. Thus Bush, who has been described as having been "prepared to be an entrepreneur of organization," conceived of a new federal agency, the National Defense Research Committee. The NDRC was approved, established, and funded directly by President Roosevelt in June 1940.[4]

With Bush as its visionary chair, the NDRC appeared to be a powerful force, but its authority ended with research on new military devices, and the president's emergency funding was soon found to be limiting. Bush, whom *Time* magazine called the "general of physics" and *The New York Times* would crown a "czar of research," then conceived of a reconstructed NDRC, one that had a direct congressional appropriation and the authority not only to carry out research on new weapons but also to produce and test them. The new agency was the Office of Scientific Research and Development. About 90 percent of the budget of OSRD was to go to funding R&D in academic settings, and about a third of that went to the Radiation Laboratory at MIT, where radar was developed during the war. Furthermore, the vast majority of the contracts ultimately gave the rights to any resulting patents to the contractors rather than to the public that funded the work.[5]

Such features did not escape the notice of some New Deal politicians, such as Senator Harley Kilgore of West Virginia. Not long after arriving in Washington in 1941, Kilgore began to promote alternatives for government control of science and technology, and, when the end of the war came into sight, he pushed for a national science foundation that would include central planning for science and technology and sponsorship of social science research. Kilgore believed that federal research activities should serve social and commercial purposes, and he also wanted to distribute research funds with some geographic equity. Vannevar Bush, seeing that Kilgore's approach would lead to the government directing research according to social ends, secured from President Roosevelt late in 1944 a letter requesting Bush to prepare a report on postwar science policy. The

The engineer Vannevar Bush, known as the "czar of research" because of his central role in distributing federal research funds, was featured on Time *magazine's cover of April 3, 1944. His 1945 report to the president of the United States was instrumental in promoting the linear model of R&D, in which basic scientific research leads to engineering development.*

president charged Bush with reporting on what could be done to make war secrets serve peacetime needs, what kind of medical research should be organized, what the government should do to aid research, and how scientific talent could be discovered and developed in American youth.[6]

In July 1945 Bush delivered his report to President Harry Truman, who by then had succeeded Roosevelt. The report, *Science—the Endless Frontier*, argued that scientific progress through unfettered basic research was essential to conquer disease, to provide new products and industries and jobs, and to enable the development of new weapons for national defense. "Basic scientific research is scientific capital," Bush wrote. His paradigm for research and development was a distinctly linear one, with basic research preceding applied, which in turn preceded new technology. He proposed a national research foundation that would be largely self-governing and that would include divisions of medical research, natural sciences, and national defense, among others.[7]

The debate over what form a national research foundation should take lasted five years; in the meantime, the National Institutes of Health had grown more powerful, and the Department of Defense and the Atomic Energy Commission had been created. Also, components of the armed services, especially in the form of the Office of Naval Research, were themselves now sponsoring more and more basic scientific research in universities. What finally did come out of Bush's recommendations and machinations, and the tug-of-war with Kilgore, was the National Science Foundation—and a model for scientific and engineering research that put research before development in name, status, fact, and deed. Engineering and technology were viewed as mere applied science, if that, and it was no accident that the enterprise of research and development, R&D, was referred to in that order. It was—and is—a grossly simplistic view of a science and technology hierarchy that carried over the typical industrial terminology and also harked back to Bush's involvement with the Office of Scientific Research and Development. Bush, perhaps more than anyone else, publicly defined, advocated, and promulgated the linear model of R&D, which would influence public policy and expectation for the foreseeable future. But, as we have seen, the history of science and technology does not support the Bushian model.[8]

Galileo, who might be considered a paradigmatic modern scientist, hardly seems to have followed the linear model of scientific thought. He improved on the technology of the telescope before using it to view the skies, and his resultant observations led to his

Dialogue Concerning the Two Chief World Systems—Ptolemaic and Coper-nican, a book of basic science that came to be viewed as a threat to both scientific and religious dogma. Put under house arrest for teaching Copernican doctrine, an aging Galileo returned to some more mundane studies of his youth and wrote his *Dialogues Concerning Two New Sciences,* in which practical engineering problems and failures motivated him to look into the fundamental nature of the strength of materials and the motion of bodies.[9]

Galileo's two new sciences are the basic engineering sciences that today's students know as strength of materials and dynamics. Each was motivated not by abstract but by practical curiosity. In the opening pages of the First Day of *Two New Sciences,* Galileo discussed explicitly, through his pedagogic characters, what Renaissance engineers did and did not know about the design and behavior of ships and other structures, which had already been developed to a high degree of sophistication. He cited a series of failures—the spontaneous fracture of stone obelisks upon being raised and of wooden ships upon being launched—as well as the erroneous engineering thinking that led to other failures, such as the mysterious fracture of a marble column just resting in storage. Galileo pointed out the existence of a size effect in structural components such as bones and noted that nature does not appear to design by geometry alone—as ship- and cathedral builders then did—but supplements geometry with some essential knowledge of strength of materials.[10]

The technology of shooting missiles and projectiles from bows, cannons, and other weapons had been developing long before Galileo set down his axiomatic treatment of motion in the Third Day of his *Two New Sciences.* Still, he no doubt benefited greatly from observing, perhaps even deliberately experimenting in thought if not in deed, before arriving at his basic science of motion. His development of two new sciences could be said to have been use-driven, that is, motivated by an incomplete understanding of the workings of technological devices and systems and thereby driven by a desire to make them work more efficiently, more reliably, and more predictably for technologists. It was development of technology that led to Galileo's mature scientific research—not the other way around.

The countless other examples from the history of science and technology include Louis Pasteur's pioneering work on the basic science of microbiology, which fed and was fed by his development of cures for disease and of methods to prevent spoiling in milk, vinegar, and wine. This and the previously discussed cases of the steam engine, powered flight, rockets, in addition to so many others, provide incontrovertible evidence for technology leading science. Basic research, in short, has long been suggested and motivated by and intertwined with technological development—and often has been led by it. Why should it have been different in the postwar years?

A close and open-minded rereading of Vannevar Bush's *Science— the Endless Frontier* reveals this actually to have been the case in Bush's time, even though the word *development* was not as tightly coupled to *research* in his report as it was in his promising paradigm. The reading of many a more recently written basic science research proposal also reveals the promise, if only subtle and implied, of a practical payoff. The National Science Foundation has long justified its very existence, as Bush did its creation, with the promise that something practical does come from basic research—ultimately. But perhaps to hedge their bets against going down blind alleys, some scientists prefer not to have to commit themselves beforehand to exactly what that practical something might be. They favor some kind of prognostication in hindsight.

In spite of, or perhaps because of, their arrogance in the wake of the Manhattan Project and under the Bushian paradigm, scientists— especially physical scientists—became the well-established arbiters of federal R&D policy in the post–World War II years. The Soviet surprise of launching the first artificial Earth satellite in 1957 was a rude awakening for America and hence for American science-and-technology policy makers. It resulted in the creation of the National Aeronautics and Space Administration and in a large infusion of new federal funds into research and development—an effort driven not by the desire to understand Sputnik in the abstract sense of pure science but with the practical and engineering goal of catching and overtaking the Soviets technologically. The Apollo program, like the Manhattan Project, was ultimately much more an engineering than a scientific endeavor. Realizing this, engineers began to seek some parity with the scientists.

In 1964, the National Academy of Engineering was created, followed in 1970 by the Institute of Medicine, both considered coequals with the National Academy of Sciences, which had naturally favored recognizing scientists. These new academies were signs that pure science, especially physics, which had been almost running amok in political influence since the technical triumph of the atomic bomb, was beginning to fall from grace. The late 1960s had brought increasing scrutiny of the R&D establishment, and Congress redefined the National Science Foundation's mission to include explicit support for the short-lived program Research Applied to National Needs, including in such areas as energy, the environment, and productivity. But RANN seemed to be a curiously unfinished acronym; it was R& without the D. Also in the 1970s, the energy crisis led to the reorganization of energy-related R&D, and growing environmental awareness presented still other new research focuses, as well as technological constraints.[11]

International economic competitiveness would come to a head in the 1980s, and then cooperation in R&D among government, industry, and universities would begin to get increased focus. Significantly, it was in the mid-1980s that the title of the National Science Board's important statistical reports, *Science Indicators*, was changed to *Science and Engineering Indicators*. In the 1990s, domestic budgetary issues would not only precipitate further fundamental thinking about R&D but would also provide a climate in which drastic restructuring would be imagined without absolute fear of letting down national defenses. Furthermore, in a period of overall federal R&D budgets that in constant dollars were flat at best, increasing numbers of scientists and engineers would be seeking a share. And there would be growing calls to rename the National Science Foundation the National Science and Engineering Foundation. One could reasonably expect a similar name change for the governing National Science Board, which first convened in 1951. Neither name was changed.

The scientists fought and won the battle of the names, but the handwriting had long been on the wall that science qua science was poised to lose at least some of its luster. The lesson had been learned at Westinghouse, where the tensions between engineering and science erupted well before World War II. A fascinating study of the competing research traditions in that corporation concluded, to the dismay of executives who had embraced the linear model after the war, that science promised more than it delivered. Typically, "research in the engineering disciplines proved to be more important to the development of new commercial technologies than even the most advanced basic research."[12]

Among those who had come to question the linear model was the Department of Defense, which in the two decades after the war sponsored scientific research in the amount of about $10 billion, with maybe 25 percent going to "basic or undirected research." To assess the "technological value" of its investment, the department sponsored a study termed Project Hindsight, which "analyzed the key contributions which had made possible the development of the twenty weapons systems that constituted, in large part, the core of the nation's defense arsenal." In 1967, a preliminary report on the eight-year study revealed that of all contributing "events," fully 91 percent were classified as technological, but only 9 percent were scientific, and of the latter, only 0.3 percent—or a total of two events—could be attributed to basic, undirected science. In other words, any "immediate, direct influence" of basic scientific research on technological development was small, if not minuscule. In 1973, Congress passed the Mansfield Amendment, which specified that federally funded research for national defense had to be correlated with military applications. This effectively meant that the linear model was no longer operative and that "practical outcomes" would dictate R&D funding priorities.[13]

In the meantime, even Vannevar Bush had come to rethink and lose confidence in his own linear model. In a 1965 book anthologizing the thoughts of engineering leaders, Bush admitted that there was a need for clarification of the relationship between engineering and science and acknowledged that "in all associations between engineers and scientists, engineering is more a partner than a child of science." He cited the development and production of the atomic bomb as a "classical example" of the partners working together. It illustrated, he emphasized, that "when science becomes applied to great objectives, so that the problems of organization, costs, scheduling, become paramount, we have then to deal with engineering." Bush also quoted Gordon Brown, then dean of engineering at MIT, whom he recalled once saying that "an engineer is a worldly scientist." It was this view of the engineer that Bush wished would be "clearly recognized and understood."[14]

As another example of engineers working together with scientists and businessmen, Bush told the story of Telstar, launched in 1962 to become the first communications satellite. He considered it "one of the finest achievements that we have seen in many years" and "primarily a great engineering job." He conceded that the groundwork for the achievement lay in fundamental science—including electromagnetism, celestial mechanics, and thermodynamics—and sound business and management practices, but he also maintained that

> engineers had to design and fit together all the elements in a working affair, to estimate costs and control performance to meet those estimates, to design so as to minimize weight, and, above all, to test every element in an elaborate system to ensure its reliability at critical moments. They had to build powerful stations in remote places, man and supply them, and tie the whole thing together so that every device and every part would operate as planned, so that nothing would be left out.[15]

Telstar certainly was a remarkable engineering achievement. But, as Bush went on to comment, "while everyone knows that engineering is concerned with the conversion of science into technology, everyone does not know that engineering also does just the opposite and translates technology into new science and mathematics." Among the examples he cited was the work of Karl Jansky, variously described as a physicist and a radio engineer, who while conducting engineering research on antennas and electromagnetic wave propagation at Bell Labs was responsible for the creation of radio astronomy. What Jansky discovered was that a source of radio noise seemed to come from the center of the Milky Way. World War II interrupted serious pursuit of this new knowledge, but wartime developments in large radar antennas and ultra-high-frequency receivers gave the "vigorous young science" a running start afterward.[16]

Beginning in the 1970s, the management-by-objectives philosophy that swept the country produced more stringent controls over industrial research and less freedom to pursue basic problems. As laboratories came under scrutiny, their staffs and their assignments were reassessed. At Corning, according to a company history of its research and development tradition, "researchers who had been wooed for their originality, as valuable high-status individual performers, now found themselves stigmatized as self-indulgent individualists who were not team players." From the late-1960s level of 25 percent, "the research portion of the technical budget was steadily reduced in favor of development and engineering and the pressure was there to reduce it still more."[17] But, overall, industrial R&D budgets grew.

By the end of the 1970s, research and development funding by private industry surpassed that by the federal government. The Bayh-Dole Act of 1980 provided a further incentive for nongovernment R&D funding to increase. The act permitted researchers working on federal projects to file for and retain patent protection rights on inventions made at universities, small businesses, and nonprofit organizations. Researchers and their home institutions were encouraged to invest their own money and other resources into commercializing new technologies that traced their development back to work done on government grants. Since 1980, the gap between private and government funding of R&D has continued to widen.[18]

In such a new climate, Vannevar Bush's linear model of research leading ultimately to development could appear not only to be overly simplistic but also to be downright backward. Public policy, if not the realities of how the world of science and technology actually works, was forcing a rethinking of how and why to allocate funds. Indeed, it became more and more difficult to defend the paradigm of unfettered pure scientific research leading to beneficial applications and solutions, and increasingly the purpose of development was seen to be justifying research expenditures. Thus research and development budgets, which in some cases in the past might have been described as "research and development of research" or "research and development of more research" or even "research and development of reasons for further research" might be said to have been self-serving. Research and development came to be seen more and more as properly being development and research. (Before World War II, this terminology was actually used for the technical department at Corning. In the early twenty-first century, that corporation was using the term "research, development and engineering," to emphasize that the task does not end until a new manufacturing process is up and operating full-scale on the factory floor.) Thus R&D might be said to have become D&R, or perhaps even "development and research for development." This last designation suggests best the continuous feedback loop that we find in real-world-directed research to solve real-world problems. It is not just that the pendulum can be seen swinging to its other extreme; it is as if it were a Foucault pendulum swinging about an axis that itself was changing.[19]

One manifestation of this was the National Research Council's proposal for a Federal Science and Technology Budget. The FS&T concept blurred distinctions between basic and applied science and technology. Furthermore, in contrast to the customary system of allocating funds, in which the total federal R&D budget was the sum of the R&D budgets of the individual departments and agencies, the proposed FS&T budget was to be constructed by first defining areas of increased and reduced emphasis, and then allocating more or fewer funds as deemed sufficient to serve not individual department and agency objectives but national priorities. A "world-class scientific and technical enterprise" would be expected to follow in areas of choice, not chance. In other words, the development of research areas would be by design—a distinctly engineering concept, activity, and perspective. Under the proposal, the several traditional research and development budget pies would be transformed into a single kind of development and research layer cake. Getting a slice of it would have to be justified by bringing to the table not just an empty plate and a sweet tooth for research funds but also evidence that one had done one's homework and deserved one's just desserts.[20]

The evolution to, or rather the return to, support for science and technology led by social and national needs meant that funding seekers could no longer successfully propose simply to do what had not been done or what could be done. It would no longer be sufficient to propose to do the next logical thing in basic research to advance or test a theory, to refine a theoretical model, or to pursue a new line of thinking with the vague promise of results sometime down the line being applicable to some unspecified practical end. Development and research had to begin with a clear articulation of the developmental problem and justify any research in terms of it.

The increasing desire of funding agencies to see multidisciplinary, multi-institutional, multi-investigator proposals, with ad hoc teams assembled to address an objective rather than to exploit a long-standing capability or maintain an existing team, was a further manifestation of the strengthening and maturing D&R philosophy. That this was, in fact, toward the end of the twentieth century a concept already somewhat in place could be seen in the creation of National Science Foundation engineering research centers, which were intended to be development-driven and finite in their federal funding life. Pure research may be for posterity, but development has an earlier shipping date.

The realities of D&R (and also by extension D&R&D) also began to be felt in graduate science and engineering education, where Ph.D.s historically have been produced in the image of their professors, a practice that goes back at least to the same World War II roots as the Bush paradigm. Indeed, one of the concerns addressed in *Science—the Endless Frontier* was how to replenish and ensure a steady supply of investigators (that is, research-university faculty and research-institute staff) to carry out the postwar investigator-driven R&D agenda.[21]

The realities of the job market forced a realization that there had been a growing oversupply of Ph.D.s for traditional placement channels, and universities were forced to respond by cutting back on the number of Ph.D. students and by altering courses, curricula, and expectations to prepare graduates for jobs in industry and other alternative markets. As the aging generation of scientists, who transformed the face of research and development funding during and after World War II, passed on, new voices of influence could be expected to continue to change the way science and technology could be expected to be supported. Where there once was a surfeit of research-ready graduates in the United States, early in the new century there came to be a paucity, caused at least in part by newly minted Ph.D.s taking their expertise back to their native countries and their own fledgling R&D and D&R programs.

The "endless frontier" metaphor, along with its implied linear research and development paradigm, was approaching the limits of its influence and the end of its effectiveness. The terminology of research and development—of R&D—that slips reflexively from our tongues may have been simply an accident of euphony or an inadvertent and unconscious act of appeasement to scientists conscripted to do engineering work in industry and later for the war effort. Regardless, Bush's linear model of research greatly affected our nation's conscious and unconscious view of the federally funded science-and-technology enterprise in its formative years. The rethinking of what the enterprise is really all about, in the light of unprecedented budget realities and by new generations of advisors and staffers who hold no allegiance to old paradigms of science, technology, or national defense, forces the R&D community to look in the mirror. Those who are honest with themselves see in that mirror not R&D but D&R.

This is nothing wholly new. The chemical engineer George Holbrook, who in the late 1950s rose to the position of vice president at DuPont, felt it "constructive to recognize that science and engineering are two parts of a single effort, each contributing its own strengths and insights, and not infrequently venturing unchallenged into each other's territory." He thought of his field of chemical engineering as "the process of blending chemistry, mathematics, physics, engineering, and economics into the solution of problems for the purpose of achieving useful results."[22]

There were no firm lines of demarcation when Corning found itself needing to make a transition from a craft tradition, in which small individual pots of molten glass were worked by hand, to a higher-volume production process. Early research provided understanding of the nature of melting and of refractory materials; later research concentrated on the forming methods. Resulting "developments in turn gave fresh impetus to glass composition research" as new problems arose in the scaling up from pot to tank melting. With this, "science was called on to find compositions that would not quickly destroy the refractory linings of what had become a very expensive item to build and maintain. The interaction between material and process was an ongoing matter, itself a major source of innovation." In other words: research led to development, which led to new research. Both R&D and D&R are really linked segments of a long and continuing line of interdependent activities and results. Perhaps we should speak of R&D&R or D&R&D, or even longer strings of *D*'s and *R*'s and *&*'s, as if they were parts of an industrial genome. It is no wonder that the shorthand R&D is used.[23]

Although the direction and funding of research at universities, national laboratories, and other institutions that were once almost totally and still are heavily dependent on government support may have been the most visible and vocal aspect of the debates that took place in the latter half of the last century, industrial sponsors of research and development also had to confront similar questions about investment in R&D and assess the results realized. Barry Allen concluded that by century's end

> the distinction between what economists call growth, what technologists call R&D, and what scientists call science was difficult to draw. There were by then more than twelve thousand large industrial research laboratories worldwide, and their work was quite as good as the best university-supported basic research, with Nobel Prizes for work at IBM and Bell Labs. Scientific instruments, material, and laboratories became industrial capital goods, and the research agenda was conditioned by the economic and geopolitical interests of the industrialized economy.[24]

At the same time, countries like China and India were becoming increasingly significant players in the worlds of education and research and development. Universities outside the United States were training a growing number of engineers and scientists, but the United States continued to maintain a large lead in R&D. A recent report by the RAND Corporation summarized the state of American science and technology in the early twenty-first century: "The US accounts for 40 per cent of global spending on scientific R&D and 38 per cent of all patented inventions among industrialised

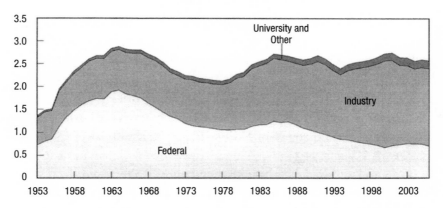

Since about the 1970s, research and development expenditures in the United States as a percentage of gross domestic product have increased from industry as government funding of R&D has declined.

nations. . . . Three-quarters of the world's leading universities are in the US and 70 per cent of the world's Nobel prize winners work there." (In 2007, U.S. industry spent about $220 billion on R&D, which was about 70 percent of the nation's total investment in science and technology.)[25]

Another report, from the National Academies—the collective name for the National Academy of Sciences, the National Academy of Engineering, the Institute of Medicine, and the National Research Council—noted that, even before the age of information technology, it was technological change that accounted for perhaps 85 percent of per capita income growth. Although the relationship is not linear, innovation (change) is still driven by research and development. But R&D budgets are driven by politics and corporate boards. Federal funding of physical science research was cut by almost half from 1976 to 2004, and in industry over about the same period the trend was away from basic research. By 2008 the vast majority of high-tech companies were cutting back on research laboratories, and the nature of work carried out in them was continuing to move away from fundamental science and toward product development. There were exceptions. Corning was putting about 10 percent of its annual net sales of almost $6 billion into "research, development and engineering expenses." Microsoft was said to be "probably the sole remaining corporate research lab that still values basic research," and yet even in that corporation, research scientists—"free-rein thinkers" working in "a pure research department reminiscent of the old Bell Laboratories"—were "far outnumbered by the thousands of Microsoft engineers working in advanced development and direct product development." Another firm that was continuing to maintain a high level of R&D was Google. It is perhaps understandable that the information technology industry would most appreciate the value of research and development as well as of development and research.[26]

The economic downturn that became so evident in the latter part of 2008 added a further wrinkle to the R&D dilemma. In early 2009, the economic stimulus package gave supporters of basic research a lift by adding tens of billions of dollars to the federal R&D budget, thus representing a "dramatic turnaround from the flat or declining research funding trends" of previous years. Unfortunately, the money came with the challenge of spending it "quickly and wisely," which was reminiscent of NASA's call for "faster, cheaper, better" missions, not all of which were successful.[27]

The many changing perspectives on R&D have important implications for approaching and solving problems that threaten the Earth. When the research community first became sensitized to phenomena associated with ozone holes, greenhouse gases, climate change, and other global issues, engineering seemed to take a backseat to science—but that could not last. The promise for solutions lay not only in the new but also in the old. According to an opinion piece in *The Economist*, the interrelated issues of climate change and energy use are also attracting the interest of more traditional companies. General Electric, which the British magazine identified as a "large American engineering firm," was reported to already have a "thriving wind-turbine business and is gearing up its solar-energy business. The energy researchers at its laboratories in Schenectady, New York, enjoy much of the intellectual freedom associated with start-up firms, combined with a secure supply of money." That would appear to have been a winning combination of engineering and economics for pressing global problems like alternative energy sources.[28]

But corporations and governments alike have only so much money to set aside for tackling all of their own and the world's problems. Choices will have to be made about where and how best to spend limited resources. The Copenhagen Consensus, a research center affiliated with the Copenhagen Business School, commissions studies to help "improve the prioritizing between various efforts to mitigate the consequences of the world's biggest challenges." Among the center's studies was one that called upon "eight of the world's top economists to identify the global challenges that can be solved most cost-effectively." According to Bjørn Lomborg, director of the center, if you had $10 billion to spend over the next four years, you should want to think carefully about how best to spend it. In order to get the best return on investment, advises Lomborg, it is necessary to weigh costs and benefits for all available options, being careful not to focus only on those that are currently fashionable. According to Lomborg's analysis, "spending an extra dollar cutting CO_2 to combat climate change generates less than one dollar of good, even when we add up all the economic and environmental benefits." He advises looking at spending money on research and development into "cleaner energy technology," which he estimates will ultimately produce eleven dollars of global economic good for every dollar spent.[29]

Philanthropic organizations, industries, and governments are faced not only with deciding whether to fund research and development but also with determining what kind at what level and on what terms. Money always seems to be scarcer than ideas, and so those with the money have to make tough decisions. This is the dilemma of the venture capitalist, who is inundated with proposals whose only sure thing is that they will cost money to fund.

The basic science of fuel cells, for example, has been known for over a century, but that has not been at all sufficient to lead the way to mass-producing an efficiently functioning cell. It behooves those who control the allocation of increasingly scarce R&D dollars to give careful thought to how much emphasis should be placed on research and how much on development. The right proportions will naturally vary from problem to problem, but in virtually all cases it will be better to err on the side of giving more to development than to research.

If development-focused efforts hit a scientific obstacle, funding emphasis can always be redirected. But if the budget is front-loaded with dollars for undirected basic research, all that may be produced is knowledge that is irrelevant and inefficacious as far as solving the problem at hand. There is certainly nothing wrong with scientists pursuing basic research in search of basic knowledge, but it is not necessarily the way to spend money allocated for attacking a particular problem. If we wish to solve real and pressing problems, we should focus on the development end of R&D.

9

Alternative Energies

When the peaceful use of atomic energy was in its infancy, scientists and engineers were very optimistic about its future. Electricity from nuclear power plants would be "too cheap to meter" one day, some believed. An "atoms for peace" program was instituted to assure the international community that the world was not ending—or at least that the United States did not intend to end it. Besides nuclear power plants, other potential peaceful applications—presented in an atmosphere that minimized, if not ignored, risk—included using atomic bombs to excavate harbors and assist in other large earthmoving projects, which would have been a great advantage over the conventional steam shovel methods used to dig a water route like the Panama Canal, a task accomplished at the expense of the great human toll taken by tropical diseases. It was foreseen that small nuclear reactors would one day power rockets and airplanes. In spite of their possible negative fallout, the beneficial by-products of the Manhattan Project salved the conscience of some scientists, who "knew the world would not be the same." One of them, the head of the project, J. Robert Oppenheimer, in reflecting on the explosion of the first bomb, at Alamogordo, New Mexico, recalled a line from Hindu scripture: "I am become Death, the destroyer of worlds."[1]

Atomic energy has been used with great success in the U.S. nuclear navy, whose power plants demonstrated the feasibility of using reactors to generate electricity for civilian use also. New technologies are always expensive and uncertain, however, and the commercial electric power industry balked at adopting the alternative to fossil-fuel-fired plants. It took government incentives to get utilities to order the first nuclear power plants. Then there was the question of risk and liability on the part of the utilities. What if a nuclear plant malfunctioned, resulting in the precipitation of radioactive materials downwind—or worse? The Price-Anderson Act of 1957 (renewed several times since) capped liability for what was then a largely untested risk and so cleared the way for the widespread commercial development of nuclear power in the United States.

Once it had gotten started, the nuclear power industry appeared to have a bright future, with few speed bumps in sight on the road to growth. One estimate made in the mid-1960s predicted that by the year 2000 nuclear power would be generating more than half of the electricity being used in the United States; this proved to be an over-estimate by a factor of two or three. In its early days, nuclear power came under the authority of the U.S. Atomic Energy Commission, which both promoted and regulated the industry. But in the 1970s the political landscape changed with the growing awareness of the environmental impacts of technologies of all kinds and the seeming conflicts of interest within the AEC. In the wake of the energy crisis of 1973, legislation was passed that broke up the AEC into the Energy Research and Development Administration, whose responsibilities included the nuclear weapons and the naval nuclear power programs, and the Nuclear Regulatory Commission, which concerned itself with the commercial nuclear power industry. Shortly after it was created, ERDA was combined with the Federal Energy Administration to form the Department of Energy. At about the same time, the Environmental Protection Agency was also created.[2]

Increasingly complex regulatory obstacles were placed in the way of nuclear power. Any proposed new plant had to contend with a bureaucratic licensing process, which involved the preparation of license applications and environmental impact statements that took up longer and longer stretches of library shelf space. The adversarial process became increasingly drawn out and promised no certain outcome. All of this added money and time, which of course is also money, to the cost of designing, building, and operating a nuclear power plant. The 1979 accident at Three Mile Island—even though it resulted in no significant release of radioactive material—completely changed the playing field and had a chilling effect on the industry. New plants that were in the planning stages were put on hold and many were canceled. The effect was to last for three decades, sending the country into an energy tailspin from which it has not yet recovered.

Few technologies grow without setbacks, and energy is a classic category with lessons for its present and its future in its past. For millennia, human and animal power met the modest needs of nascent societies. In more mature civilizations, great efforts of organization were called upon when necessary or desired to construct monuments like the Egyptian pyramids at which we still marvel. Water and wind power, among the cleanest, most inexpensive and renewable of all sources, drove mills for centuries, with clever engineering making them more powerful and efficient and devising schemes to make them more versatile. The world might still operate on water and wind, were it not for the desire to have power where there was no water flowing or when there was no wind blowing. The steam engine, and later the internal combustion engine, addressed this limitation to such an extent that water- and windmills became quaint and evocative elements in pastoral landscapes painted by nostalgic artists. In time in America, old mills of all kinds, and their machinery, became neglected and abandoned as the textile and other industries moved south—where there was relatively inexpensive electric power and an unorganized unskilled labor force—and, still later, offshore.

With the widespread introduction of farm tractors and the electrification of rural America in the early part of the twentieth century, power had become portable and personal. As long as oil flowed like water and electricity like light, the future looked bright. Unfortunately, in the 1970s, political skirmishes in the Middle East caused oil to become viscous in the pipelines whenever the chill of a crisis developed. Electricity, in some cases directly dependent upon oil, also had its interruptions, including massive power failures that revealed the fragility of regional networks and therefore of the national grid. Engineers knew about this and had done their best to design the systems to accommodate the occasional malfunctioning switch and burnt-out transformer, but even engineers cannot always anticipate everything—or ensure a system's constant operation in the face of increasing demands.

The vulnerability of the nation to an oil or power crisis, coupled with the heightened awareness of the financial and social cost of energy and of overarching environmental issues, made for a volatile and uncertain mix of resources, methods, and issues. With nuclear power effectively off the table in the wake of Three Mile Island, there came increasing calls for research and development into alternative sources of energy, by which were meant principally wind and solar power. (Energy from water, embodied in some massive hydroelectric power plants and exploited on a large scale since the early part of the century in such grand projects as Hoover Dam, was no longer fashionable, since there had been growing concern that dams changed the ecology of a region. Additionally, the most logical and desirable locations for hydroelectric plants had long since been developed.)

Windmills, renamed wind turbines, began to reappear on the American landscape, as they did elsewhere around the world. In the Netherlands, long associated with the picturesque broad-bladed windmill, sleek modern wind turbines on steel towers now dwarf their predecessors in size and number. In America, where a windmill once stood atop a tall tapered steel or wooden tower erected beside virtually every barn and farmhouse, now new windmills are appearing with increasing frequency. In Southern California, on the 1,500-foot-high San Gorgonio Pass near Palm Springs, thousands of wind turbines stand straight up, looking like giant soldiers twirling their rifles. They also resemble a sparse growth of tall, pale vegetation. Other so-called wind farms are beginning to proliferate around the country. And because wind farms have become big business promising big profits for early (and often subsidized) investors, the rush to install electricity- and noise-generating turbines can produce undesirable social side effects in economically depressed communities, where some citizens are all too eager to share in the windfall. This was the case in rural upstate New York, where neighbors did not agree on which way they wished the winds of change to blow. Among the charges of "corruption and intimidation" levied by some citizens were conflicts of interest among town officials who were voting on regulating the machines and at the same time taking money to allow them to be installed on their land in a profiteering reversal of the not-in-my-backyard syndrome.[3]

Collections of wind turbines planted like crops in a field have come to be known as wind farms. The large-scale generation of electricity from the wind, which frees up natural gas for other uses, has been a central component of the energy independence plan promoted by the oilman T. Boone Pickens.

Among the most widely publicized opposition to the installation of wind turbines has been that to the Cape Wind project in the waters off Nantucket Island. Residents of this celebrated summer retreat of affluent and influential people, many of whom also might identify themselves as environmentalists, raised vocal opposition to the installation of 130 wind turbines that they claimed would ruin views and injure birds. Locating wind farms even farther offshore, where the wind tends to blow more reliably, also makes a lot of sense technically. Experience gained in installing and operating oil and gas rigs in deep water in the Gulf of Mexico and elsewhere is directly transferable. Whereas many states have begun to explore establishing offshore wind farms, New Jersey has been most aggressive. It set a goal of generating three thousand megawatts of wind power by the year 2020. That would be the equivalent of about three nuclear power plants and would represent about 13 percent of the state's total power generation—enough to supply almost a million homes with clean, renewable electricity. Some of the wind turbines would be located as far as twenty miles off the coast near Atlantic City and were promised to "appear to be half the size of your thumbnail and as thin as a toothpick."[4]

Wind farms could be located in remote areas, where few if any people would see them at all. One such location might be around the sparsely populated Upper Peninsula of Michigan. But generating power far from population centers would necessitate the construction of long-distance, high-voltage transmission lines. The installation of such lines not only is expensive, thus increasing the capital investment required, but also is liable to raise objections from people through whose area the lines must pass. There are no easy solutions to tough problems.[5]

While the prospect of windmills elicited outrage from Nantucket Islanders and brought corruption to a bucolic New York community, a proposal to employ them throughout the crowded boroughs of New York City brought confusion. At a National Clean Energy Summit in Las Vegas, Mayor Michael Bloomberg proposed installing wind turbines on the tops of New York's tall buildings and bridges and in the waters off the boroughs of Brooklyn and Queens. Almost immediately, architects and engineers questioned the idea, citing both aesthetic and structural objections to its feasibility, not to mention wondering how much power could be produced in a city not noted for its wind. Not only might the devices deface historic sites like the Brooklyn Bridge, but also existing structures were not designed to take the twisting, bending, and vibrating forces that would accompany the installations. That is not to say that new large bridges and buildings could not be designed to accommodate such forces. An early concept for the World Trade Center's Freedom Tower incorporated enough wind turbines into the upper reaches of its structure to operate its elevators, or so it was hoped. Other buildings being designed for sites around the world increasingly have similar green energy schemes associated with them. However, it has been suggested that the energy consumed in the manufacture, transportation, and installation of turbines might in fact produce "more carbon dioxide emissions than the turbines save." In the final analysis, it is a question of whether the power produced provides a suitable return on the tangible and intangible investment.[6]

By mid-2008, before the financial crisis hit full force, the "wind industry" was expanding at such a brisk pace that there was a worldwide scarcity of the essential turbines. Those that were installed were generating about 1 percent of the electricity flowing through the U.S. grid. (At that time, fossil fuels were generating about 70 percent, nuclear about 21 percent, hydroelectric about 7 percent, and renewable energy sources, including wind and solar, about 2 percent.) A report by the Department of Energy projected that the contribution from wind turbines could grow to 20 percent by 2030. This would not only reduce (percentage-wise) our dependency on fossil fuels but also would reduce emissions of carbon dioxide from fossil-fuel plants, as well as lower the consumption of water, which is used to cool those plants. The escalating price of oil and natural gas made wind farms attractive investments, which drove the positive projection of their growth. But the credit crisis, coupled with falling stock prices, created a dramatically different climate for alternative energy companies, causing them to have a difficult time raising capital to forge ahead with their development efforts. This, coupled with declining prices for oil and natural gas, resulted in many ambitious alternative energy projects being put on hold. These included bio-fuel, solar, wind, and electric car initiatives.[7]

Regardless of the financial climate, for every energy advantage there is a corresponding disadvantage or impediment. Since wind does not blow on demand, and is in fact inconsistent and not totally predictable, electric-grid engineers think it is a "nightmare" to deal with. In order to incorporate wind-generated power into the grid on a large scale, it will be necessary to modify the grid so that it can accommodate a wide variety of power sources, including less environmentally friendly ones to back up those that are friendly but fickle. There will likely be the need for new transmission lines and "smart meters" that can handle power going both ways—from the grid to the consumer and also from the consumer who has a small wind turbine that can feed its excess power into the common grid. These are the kinds of investments that are not always considered in overly optimistic projections of more environmentally friendly power sources.[8]

Furthermore, whereas the cost of power from a large wind farm might be less than ten cents per kilowatt hour, that from a single high-tower wind turbine might be five times as much, and that from a small individual turbine could approach fifteen times as much. There is definitely an economy of scale in wind power, which explains why so many turbines tend to be clustered together in farms. In spite of the fact that an investment in a small wind turbine will not likely ever be repaid, it has become fashionable to have one's own source of green power. Small wind turbines barely capable of generating a single kilowatt have been installed on home rooftops—as spindly television antennas were in the 1950s—where they provide a "far more visible clean-energy credential than solar panels, which are often hard to see." For those who prefer a relatively low-profile unit, the Swift Wind Turbine, whose reinforced-plastic blades sweep out a seven-foot-diameter circle, needs to clear only two feet above a roofline to operate. Manufactured by Cascade Engineering of Grand Rapids, Michigan, this turbine can generate as much as 1.5 kilowatts when the wind is blowing at thirty miles per hour, which in most locations is not often.[9]

Unadorned rooftops themselves have been identified as potential sources of energy, albeit conserved rather than generated. In particular, a white or light-colored roof makes for a cooler building beneath, lowering air-conditioning needs in warm climates and hence reducing carbon emissions. Since 2005, it has been the law in California that flat roofs, which cover mainly industrial and commercial buildings, have to be white. And, since the summer of 2009, even sloping roofs put on new residential buildings have had to be "light-colored cool-roof colors, if not white." Such mandates were motivated by the fact that "a 1,000 square foot area of rooftop painted white has about the same one-time impact on global warming as cutting 10 tons of carbon dioxide emissions." This is so because of the amount of sunlight reflected back into the sky rather than absorbed by the roof.[10]

Whether installed on a rooftop or on the desert floor, solar-panel energy has been notoriously expensive, largely because of the high costs of producing the photovoltaic cells that convert the energy in the sun's rays into electricity. Still, solar cells have been called the "ultimate green technology," and there has been considerable interest in finding a means of mass-producing them to sell at a price competitive with traditional sources of electricity. One of the principal factors keeping the price of solar cells high is that the vast majority of them require for their production wafers of pure silicon. But these are also key building blocks for the enormous semiconductor industry, with which solar energy producers have had to compete for raw materials. This is why promising alternatives to silicon are looked upon so favorably by solar cell manufacturers and are pursued so aggressively by start-up companies and venture capitalists alike. In one process, a nanoparticle ink containing copper, indium, gallium, and selenium atoms in appropriate proportions can print solar cells on rolls of plastic or aluminum foil as if printing a newspaper. Projections that this could result in solar cells that might be sold for one dollar per watt—about one-third the cost of silicon-wafer-based cells—have made future solar energy appear to be competitive with electricity generated by coal, gas, and nuclear power without the need for subsidies, which have proven to be problematic. (Nevertheless, among the features of the economic bailout package passed by the U.S. Congress in the fall of 2008 was a provision for homeowners that extended for eight years a 30 percent tax credit for installing solar panels.)[11]

But until their price does become competitive, costly solar panels will be the object of desire for more than environmentalists. Just as aluminum siding and copper wire and tubing from plumbing and air-conditioning units have been stripped from houses and sold when the prices of those metals made the effort and risk worth it, so thieves have found the high cost of solar panels something to trade in. California, which has tens of thousands of installations, has been the scene of solar panel burglaries. One thief stole some panels from a toll road and offered them on eBay. Astute Newport Beach police were the high bidder on the lot, which was worth about $1,500. When the culprit came to deliver the merchandise, he was arrested. His defense was that he bought them on craigslist and had no idea that they were stolen. Thievery of panels is said to be entrenched in Europe, where the solar industry has been well established.[12]

Germany, known for being environmentally minded, in 2008 was generating 14.2 percent of its electricity from renewable energy sources. Yet that country, which receives yearly only about half of the sunshine that San Diego sees, and which was producing only 0.6 percent of its electricity from the sun's rays, rose to be a leader in the solar cell industry, thanks to significant government subsidies, financed by a surcharge added to monthly electric bills. By law, the surcharge was to increase with the growth of the industry, but because it began to grow so fast consumers were expected to react negatively and oppose continued government subsidies. The chief executive of Q-Cells, the German company that had risen to be the largest supplier of photovoltaic solar cells in the world, argued for continued support, saying that "to develop a technology, you've got to create an industry. You can wait and wait and wait for costs to come down, but it takes too long." Such are the realities of the world of technology and business. Engineers can do only so much in bringing a product, especially an innovative one, to market at a reasonable price.[13]

The town council of the German city of Marburg got so caught up in a desire to promote renewable energy that it changed a policy of encouraging to one of requiring the installation of solar panels. The policy applied not only to new construction but also to any house in which a new heating system was to be installed or on which roof repairs were to be made. The fine for not complying was set at €1,000, or about $1,500 at the time. The ordinance threw the citizens of Marburg into heated debate over where to draw the line on "ecological good citizenship" and led to accusations that their town government had developed into a "green dictatorship." One citizen, who had wished to reinsulate his home to make it more energy-efficient, did not do so because he would have had to spend an additional $8,000 to install solar panels. He considered the town council's mandate to be "absolutely counterproductive."[14]

Environmental concerns and politics always have to be factored into the energy equation. Among the most attractive locations for large solar power plants in the United States are in the Southwest, where there are vast swaths of sun-soaked public land. The Bureau of Land Management had been considering proposals from solar energy firms to establish large plants on such lands, but the great number of applications—potentially affecting over a million acres—led in the summer of 2008 to a moratorium on new projects until their environmental impact could be assessed. Among the concerns was the effect of the plants and associated transmission lines on native wildlife, such as the desert tortoise and Mojave ground squirrel. There was also concern about water use by solar power plants, which have to employ the scarce desert resource to wash dust off the panels or to generate steam in plants where myriad mirrors concentrate the heat of the sun on a boiler. Since investment tax credits were set to expire before the environmental impact study was expected to be completed, solar energy businesses were having to consider the possibility of using private land for their installations, adding considerable expense to the enterprise. Investors, developers, and speculators protested so loudly that the bureau reversed its decision and accepted new applications. No development project is strictly about science and engineering. There are always complicating issues, including aesthetics, economics, ethics, and politics.[15]

In 2002, California mandated that by 2010 utilities like Pacific Gas & Electric had to provide 20 percent of their electric power from renewable resources. (By 2008, half the states were requiring utilities to include renewable energy in their portfolios.) In an effort to meet the requirement in California, two enormous solar power plants were being planned. Together they were to cover more than twelve square miles with solar panels capable of generating eight hundred megawatts (almost as much electricity as a typical nuclear or fossil-fuel plant) when the sun is bright, but since they cannot operate continuously like a coal or nuclear plant, they might produce only a third as much total electricity. One of the solar plants will be built by Opti-Solar, a company that will employ "a thin film of active material" on its panels. The other plant, to be built by SunPower, will use a technology based on silicon crystals. These installations were to represent a twelvefold increase in rated capacity over the largest plant operating in 2008. According to the chief executive of OptiSolar, "If you're going to make a difference, you've got to do it big."[16]

Sunny California has understandably been far ahead of all other states in installing megawatts of solar power, but just having abundant sunshine is not enough to promote a widespread solar economy. Nevada, said to have "some of the best solar potential in the world— enough to power the entire nation," has neither enforced mandates nor offered incentives to exploit its natural resource and so has not been anywhere in the running. But some energy-conscious enterprises in the state have found their own incentives for installing solar panels. Two nonenergy firms erected solar panels on an elevated framework above their common parking lot so that the installation not only provides about half the power consumed by the firms but also shades employees' cars from the hot desert sun.[17]

Second place behind California in installed solar capacity was occupied by a surprise competitor—New Jersey. The Garden State achieved this distinction through a rebate program so attractive that the system choked on its own success. An average homeowner could receive a rebate of $20,000 and a large commercial installation as much as $1 million. Applications for rebates backed up in the bureaucratic pipeline and drained the state's rebate account, which was funded by a surcharge on utility bills. It became clear that the time had come to rethink public subsidies for solar energy. The subsidies would be replaced with energy credits that could be traded on the open market. Credits would be earned by solar projects, and utility companies would be required to buy them to offset their plants' carbon emissions and thereby contribute to meeting targets for renewable energy. Eventually, all renewable technologies will have to compete "in the wild," however, without being clothed in the attractive financial incentives that made New Jersey number two.[18]

The idea of energy credits traded on "pollution markets" was introduced in the United States during the period of environmental consciousness-raising of the 1970s. When the concept was applied in the 1990s, it had some effect on curbing acid rain. According to one report, the recent "carbon markets," also called cap-and-trade systems, have become popular because they are more politically palatable than imposing taxes on carbon emissions. The scheme involves placing limits on greenhouse gases; utilities then can buy or sell permits, depending on whether they exceed or come in under their limits. Unfortunately, there is no guarantee that there will be a global benefit. The largest market for greenhouse gases in the world was created in Europe in 2005, and the early results have been disappointing at best. In many industries carbon dioxide emissions continue to rise rather than fall. Financial incentives and disincentives may promise to work from a regulatory point of view, but they alone will not reduce carbon emissions; nor can they improve the efficiency of solar panels or make them cheaper to manufacture and install. Achieving those goals will require concerted collaborative efforts in science and engineering.[19]

In addition to wind and solar, geothermal energy promises to be another clean alternative to fossil fuels. Indeed, it has been described as "perhaps the most under-appreciated potential source of electrical power." Here and there around the surface of the Earth, geothermal energy manifests itself in hot springs and geysers, but it is ubiquitous within the interior of the planet. The heat trapped inside the Earth can be tapped for directly heating buildings or for indirectly driving electrical generators. According to a 2006 report issued by the National Renewable Energy Laboratory, in the United States there is "enough geothermal energy within two miles of the surface to supply the nation with 30,000 years of energy at current rates of consumption." A subsequent report by the Geothermal Energy Association described geothermal energy as "an underestimated, under-reported, under-explored, and under-studied natural resource." Accessing this energy, which can be described as "mining heat," is akin to drilling for oil, but the much greater temperatures and caustic environments encountered make for a significantly greater technological challenge. Nevertheless, harnessing this energy has proven practical. Thirty-five years after the Philippines began developing geothermal power, it was being used to generate 28 percent of that country's electricity, the largest proportional use in the world. Power derived from hydro-electric dams and underground geothermal fields in Iceland has been considered one of the country's greatest natural resources. Since such a resource is not easily exported, the government strove to "import demand." The aluminum industry, which requires great amounts of energy to operate, was thus a natural one to woo, but when the government wished to attract a third plant to the country, environmentalists protested the damage to the landscape that such a heavy industry can cause. The availability of clean energy does not guarantee the clean use of it.[20]

Renewable sources are only one side of the global energy equation; the other side relates to how electricity generated from those sources is used and, where possible, stored. While concerted efforts to improve batteries have been going on for 150 years, batteries still remain relatively heavy, expensive, and of limited capacity and life, thus making electric cars uncompetitive. As presented in an article in *Design News*, the technical logic is inescapable: "To an engineer, it looks obvious. Gasoline packs 80 times more energy per kilogram than a lithium-ion electric vehicle battery. It holds 250 times more energy than a common lead-acid battery. So, it's a no-brainer. Batteries can't possibly deliver the energy needed to power the future of the auto industry, right?"[21]

But changing conditions, attitudes, and expectations can change no-brainers into brainteasers. Tightening emissions standards, rising fuel costs, and even the possibility of fuel shortages associated with international tensions, call for some rethinking of the once obvious. According to the director of the Center for Automotive Research, what is needed to produce effective battery-powered cars is not so much invention as engineering. Much of the necessary technology is believed to exist, but sound engineering development remains to be done. There is general agreement that a large and fully equipped battery-powered car is not yet feasible, meaning that no battery pack can last for three hundred miles and be recharged in fifteen minutes—the time it takes to make a pit stop, fill a gas tank, and buy some snacks to eat in a conventional automobile.

But General Motors was so confident that its engineers, and battery suppliers, could find a mixed-mode solution that in the spring of 2008 it promised a 2010 delivery date for its lithium-ion-battery-powered "plug-in hybrid," the Chevy Volt. The Volt's batteries can be recharged overnight in the owner's garage so that it can get as much as 150 miles per gallon of gasoline and go as far as 40 miles using nothing but battery power. This, GM advertised, would mean that more than three-quarters of Americans who commute daily could do so without using any gasoline. Even though "production-ready prototypes" were ready in the summer of 2008, there was still a lot of work to be done on the battery. The exterior design of the car had already undergone changes in styling, including a shortened hood and a more rounded front end, resulting in the addition of about six or seven miles to the vehicle's battery range. Such incremental gains are typical of the engineering development process, but they take time and involve cost and risk for the company. A four-door Volt was expected to sell in the $30,000 to $40,000 range, with buyers being eligible for a tax credit of as much as $7,500. General Motors was not expected to make money on the car in the short term, but in the long term the battery technology developed was expected to enable the production of highly efficient and profitable vehicles. Within months of GM's announcement, Chrysler made one of its own. But instead of developing an entirely new car, which the Volt was to be, Chrysler would convert some of its existing model designs into electric vehicles. This would allow the company also to deliver an electric car by 2010. In the meantime, a limited number of hybrids had been converted to plug-ins.[22]

The Silicon Valley start-up electric automobile company Tesla Motors had promised to be selling a luxury car termed the Model S by mid-2011. By pricing the car in the $60,000 range, the company hoped that its expensive (and, at a thousand pounds, heavy) lithium-ion battery pack would not so dominate its production cost. The rollout of the car had to be delayed when capital became tight in 2008, slowing down engineering work on the Model S, but Tesla continued to make its Roadster, which carried a price tag of more than $100,000 and at the time had a waiting list one year long. Whether development and engineering are all that remain to produce competitive electric cars and competitive solar panels remains to be seen. Engineering may not need complete scientific knowledge to advance the state of technology, but it cannot do what nature will not allow. As Richard Feynman is remembered to have said, "Nature cannot be fooled." Whether or not engineers know her rules and laws and limitations, those are what will govern what can and cannot be known, what can and cannot be achieved. To nature's limitations must be added, of course, those of capital.[23]

Other renewable energy sources include the ocean's waves and tides, which some scientists believe to be vast unexploited resources. According to one estimate, harnessing just 0.1 percent of the ocean's energy could supply the electricity needs of fifteen billion people. Although the wind does not always blow or the sun always shine, the twice-daily tides represent a totally reliable source of power. Passamaquoddy Bay, at the mouth of the St. Croix River, which runs along the border between New Brunswick, Canada, and Maine, has the greatest tidal change in water level (about twenty feet) of any location in the continental United States. President Franklin Roosevelt, whose summer retreat on Campobello Island was not far from the largest whirlpool in the Western Hemisphere, promoted the Passamaquoddy Bay tidal power project. Like the tide mills that once drove industries throughout coastal Maine, the Roosevelt project relied on dams and impoundments that trapped water at high tide and released it at low tide to produce power. The new approach employs tidal stream turbines that, because the density of water is so much greater than that of air, can be many fewer in number than wind turbines to produce the same amount of power. And, because the stream turbines are underwater, their installation draws much less opposition attributable to ruined views or disturbed tranquillity than do massive wind turbines that tower and hum over the water. Although by some estimates deriving power from the tides is more than a decade behind extracting it from the wind, the technology is believed to be in place: "It's just a matter of engineering it for the lowest cost, the highest reliability and the longest survivability in a hostile and corrosive environment." Still, that kind of engineering does take time. Meanwhile, tidal power is definitely emerging, albeit slowly, as a technology to watch.[24]

Another reliable source of water power is the Gulf Stream, which provides a constant flow of more than eight billion gallons per second. If turbines are placed in the stream at a depth of thirty to forty feet below the surface of the water, they will be out of reach of shipping, and the power they generate can be transmitted through undersea cables. Such a scheme, employing thousands of turbines, might produce as much power as ten nuclear plants and supply one-third of the electricity needs of Florida. All such projects tend to think big but necessarily start small: exploiting the Gulf Stream began with a $5 million grant from the state of Florida, which would enable further study, including the installation of a small test turbine. Even should such a project recover from anticipated early setbacks, it will have to contend with opponents who worry about the "Cuisinart effect," in which the spinning blades might be capable of chopping up sea creatures, including fish. The wind turbine industry had to deal with the idea that it would do the same to birds.[25]

Some projects to harness the energy in the sea have had bad luck with early efforts: a $2 million test buoy was lost off Oregon's coast, and blades broke off two test turbines installed in New York City's East River. But Verdant Power, the company responsible for the New York project, believed that it could succeed by making the rotor blades out of a new material. Instead of the steel-reinforced fiber-glass blades used on the first turbine or the aluminum- and magnesium ones used on the second, Verdant made its third set of blades out of an aluminum alloy. Once it found the right material, it expected to install three hundred turbines in the river and thereby provide power to as many as ten thousand New York homes. Like a typical large-scale engineering project, this one started small, with six turbines providing electricity to a supermarket and parking lot on nearby Roosevelt Island. The company's persistence was expected to pay off because of the promising nature of the power source. Still, no clean power is without its drawbacks, and every kind of energy source has its pluses and minuses. A news report put it concisely: "Tidal power is reliable, but small-scale. Wind power is cheap but rare. Solar power is unreliable, inconsistent and expensive but easy to install."[26]

The use of water currents or waves to generate electricity has come to be known as hydrokinetics, as opposed to hydropower, which exploits the potential energy of water impounded behind a hydroelectric dam. Harnessing a river's flow to generate electricity is a century-old technology, and the state of Washington gets 70 percent of its power from dams. But not all rivers can or should be dammed. One new hydrokinetic technology known as VIVACE (an acronym for Vortex Induced Vibrations Aquatic Clean Energy) employs not stream turbines but cylinders mounted on springs. As the water passes around the submerged obstacles, vortices are created that move the cylinders up and down, a motion that can be used to generate electricity. Such a system can operate in currents of less than one knot (or about one mile per hour), which makes it more broadly applicable than turbines, which require a current five to six times as fast in order to work efficiently. Since most of the world's waterways move at less than three knots, this new technology is very promising indeed. A prototype has been installed in the Detroit River, which flows at less than two knots. The Mississippi and Ohio rivers have also been looked to for hydrokinetic installations of the more traditional turbine kind. The company Free Flow Power sought permits to install on the order of 2,500 stream turbines in the Ohio, but potential adverse environmental impact is likely to impede the project. The advantage of a system like VIVACE is that the blunt cylinders move relatively slowly and so have less chance of harming underwater wildlife.[27]

There are also more pedestrian schemes to generate clean power. British engineers have proposed harnessing the "footfall of trudging shoppers" walking through malls and other public places. These "underfloor generators" would be powered by "heel strike" and would employ foot pressure to squeeze fluid from pads embedded in the floor through miniature electricity-generating turbines. The technology had its origin in the American military, where small generators in the boots of soldiers were employed to reduce somewhat the burden of carrying heavy batteries into combat. The underfloor concept has been successfully incorporated into a prototype system installed in the offices of a London company. Similar devices could be employed on roads and bridges to harness the energy imparted to the structures by passing motor vehicles and railroad trains. However, like most new and imaginative technologies, underfloor generators are expensive to manufacture and to install, so they will likely receive limited use, at least for the time being.[28]

An alternative means of generating electricity from footfalls was incorporated into a nightclub in Rotterdam, where some ecologically motivated Dutch inventors, engineers, and investors formed a company called the Sustainable Dance Club. The dance floor in their hip Club Watt exploits the piezoelectric effect, whereby the pressure exerted by dancing feet produces electricity in the special material used in the floor. The more the patrons dance, the more electricity they produce. Like the mini-turbine technology, the piezoelectric one started out expensive, but the Dutch company expects that it will develop materials that are less expensive and more effective, thus making the system cost-efficient. Meanwhile, it was looking to employ the piezoelectric technology in gyms and fitness centers, where there is plenty of untapped pedal energy.[29]

Another nascent technology has been developed by a Georgia Tech engineering professor and his colleagues. It involves the generation of electricity simply by flexing a certain kind of wire back and forth. When properly insulated, a number of such wires could be embedded into the sole of a shoe and so produce current with each footstep. Alternatively, collections of the wires could be woven into garments, so that body movements of all kinds might produce electrical energy that could be used to power personal electronic gadgets like an iPod, a medical device, or a small computer. The wires might even be incorporated into a flag to produce a different kind of wind power. While prototypes of the so-called nanogenerator produced only a fraction of the voltage delivered by an AA battery, there was the potential for scaling up. The fate of such a development would depend greatly on how economical it would be.[30]

In the summer of 2008, when crude oil was trading near $150 a barrel and the average cost of gasoline at the pump was more than four dollars a gallon, there was a heightened interest in alternative energy sources and fuels. For years, biofuels—those derived from plant matter—had been promoted as an environmentally sound and renewable alternative to fossil fuels. In 2005, the U.S. Congress mandated a renewable fuel standard requiring that corn-based ethanol be blended into gasoline. The legislation provided for annual escalations of how much ethanol was to be used, which was blamed for causing corn prices to almost quadruple over four years. This in turn impacted cattle ranchers, chicken farmers, and the food industry generally—and ultimately the consumer—with higher prices. The legislation empowered the Environmental Protection Agency to issue waivers if the new mandates turned out to have unforeseen negative consequences. Though agency economists did not find that the use of corn-based ethanol was damaging enough to warrant a waiver, a study by the World Bank concluded that since 2002 as much as 75 percent of the increase in food prices was attributable to biofuels. According to an editorial in the *Christian Science Monitor*, this was all "for a fuel that contains one-third less energy than gasoline, reduces mileage per gallon and, for many vehicle owners, damages engines."[31]

It was not long before some other disadvantages of biofuels began to surface and be reported in the mass media. Ethanol, which had been used in small amounts (approximately 1 part per 10 parts of fuel) as an additive to gasoline, was being promoted for use in much larger concentrations as a "homegrown alternative to gasoline." One much-talked-about new fuel was E85, made by blending 85 percent corn-based ethanol with 15 percent gasoline; it could be used in internal combustion engines designed to burn pure gasoline without their having to be modified in any way. Midwestern corn-growing states, where a third of the production finds its way into ethanol, were promoting the use of E85, as was the federal government. In Minnesota and Iowa, gasoline retailers were being offered financial incentives to install E85 pumps; Iowa, which grows the most corn, had plans to triple the number of such pumps over a two-year period. Public policy had gotten ahead of engineering, however, and it was discovered that the gasoline alternative might damage metal and plastic parts used in fuel pumps. It is a familiar phenomenon that substances that are benign in small concentrations can become aggressive in larger ones, and apparently E85 was capable of corroding some metals and embrittling some plastics, potentially posing considerable maintenance, public safety, and environmental issues.[32]

Such problems can be overcome through research and development focused on identifying more resistant materials. However, there may be more fundamental environmental problems that come with the excessive promotion of biofuels. If, for example, farmers were to become overly aggressive in fertilizing their cornfields in order to produce greater yields to meet biofuel demands, it could result in a situation in which nitrogen damage outstripped the carbon benefits realized in the crops. The amelioration of global warming must be weighed against the exacerbation of other environmental concerns, including nitrogen-related smog and acid rain.

Since nitrogen is a key ingredient in fertilizers, its profligate use in agriculture could bring to the fore a hitherto downplayed factor in global climate disruption. In order to sharpen their message, over the past couple of decades environmentalists appear deliberately to have focused on carbon rather than any number of other elements as a threat to planet Earth. Now, having to deal with a host of problems, some environmentalists worry that "after the hard-fought campaign spotlighting carbon, turning to focus on nitrogen could upset the momentum." And then there are the additional "heat-trapping emissions," including chlorofluorocarbons, methane, and soot resulting from burning diesel and coal. Al Gore, the environmentalist's environmentalist, recognized the danger of communicating the complexity of the problem: "Look, I can start a talk by saying, 'There are 14 global warming pollutants, and we have a different solution for addressing each of them.' And it's true. But you start to lose people."[33]

The spring and summer of 2008 also revealed another risk associated with relying on agricultural crops as the source of a renewable fuel component in gasoline. The season's floods, which raised the price of ethanol by almost 20 percent in one month, were not expected to have a very significant impact on the price of gas at the pump, since the amount of ethanol in the total fuel mix was still relatively small. But the same could not be assumed for future years, when the contribution of corn-based ethanol to the national fuel supply was projected to be much greater. According to an energy consultant, "There is now a vulnerability to perfect storms, not just in a metaphorical sense, but increasingly in a literal sense." He believed that weather risks had to be added to geopolitical risks. A different energy analyst likened the nation's energy policy to "playing Russian roulette with every chamber loaded."[34]

Another once-promising plant-derived fuel is palm oil. In the Netherlands, government subsidies were used to entice electric companies to rework their power plants to use biofuels like palm oil, which theoretically burns cleaner than fossil fuels. But the increasing demand for palm oil in Europe led growers in Indonesia and Malaysia to clear large tracts of rain forest and use excessive amounts of chemical fertilizer. According to one reporter, what was supposed to be a "green fairy tale began to look more like an environmental nightmare." To make matters worse, in Indonesia more space for growing oil palms was created by draining and burning large fields of peat, which released so much carbon into the atmosphere that country soon became the third-highest producer of carbon emissions in the world. Since global warming is by definition a worldwide problem, if the Netherlands produced less pollution by generating electricity using palm oil that was produced in an area that was contributing more than its share of greenhouse gases, there could be no net benefit to the global environment. Alternative fuels must be beneficial in more than a local way to have a net-positive effect on the planet. There must be a world-systems engineering approach to developing biofuels to ensure that this is achieved.[35]

Systems engineering has been defined as "the invention, the design, and the integration of the whole ensemble" and has been said to be "an old and ever-present part of practical engineering." Like much engineering, it involves a lot of common sense. Who can imagine that the production of biofuels or anything else can proceed totally independent of other natural, artifactual, and social activities? According to the legendary engineer-educator Hardy Cross:

It is customary to think of engineering as a part of a trilogy, pure science, applied science, and engineering. It needs emphasis that this trilogy is only one of a triad of trilogies into which engineering fits. The first is pure science, applied science, engineering; the second is economic theory, finance, and engineering; and the third is social relations, industrial relations, engineering. Many engineering problems are as closely allied to social problems as they are to pure science.[36]

A wide variety of plants and plant products has been employed in making biofuels. Brazil uses sugar instead of corn to make ethanol. European countries use, in addition to palm oil, rapeseed and sunflower oil. Converting farmland from food to fuel production can obviously affect the supply and hence the cost of food. For this reason, using crops such as wood has been promoted, not only to preserve food crops for their originally intended use but also to revitalize depressed sectors of the economy, such as the forest products industry in Maine. In some cases, it has been proposed to use fast-growing and hardy grasses for fuel production, but this has created problems with the weeds invading nearby food-producing land, thereby increasing the cost of growing the edible crops or reducing the yield of that land. A proposal by the Michigan-based company Sequest has been to combine carbon dioxide from a coal-burning power plant with wastewater from a nearby sewage treatment plant to grow algae that could be made into biofuel. This would be a creative environmental engineering experiment in how to make good use of a greenhouse gas rather than having to sequester it.[37]

System and societal effects have included more than increases in food prices. In 2007, so many new ethanol plants were being built that their estimated completion time was being delayed by almost a year because of shortages in such seemingly common building materials as galvanized steel and more specialized components like distillery tanks. Small farming towns, concerned for their water supplies, were beginning to demand environmental impact studies for proposed ethanol plants. The rising prices of fossil fuels and corn had created a "furious gold rush" to invest in new ethanol plants. It was being predicted that ethanol plants could soon be using as much as half of the country's corn crop, and there was growing concern that the ethanol industry was expanding too fast. Economists expressed concern that the rush to produce corn-based fuel would escalate prices of livestock and retail food products, something that consumers witnessed two years later. When the economy shifted dramatically in the fall of 2008, plants still in the planning stages were put on hold. Unstable corn costs, falling ethanol prices, and tight credit combined to make a sure thing look like a pig in a poke. Systems thinkers too often are not consulted in devising planning models.[38]

. . .

Another alternative for reducing the cost of operating a car or truck is, of course, conservation. When gasoline prices reached unprecedented heights, drivers of all kinds tried to squeeze more miles out of every gallon. Some did this by using cruise control more, avoiding braking on the highway almost to the point of danger. Others did not use their air conditioners, because they had heard that they could get three more miles per gallon that way. Others, who had dashboard displays that gave them regularly updated readouts of average miles per gallon, spent a good deal of driving time studying those readouts, trying to learn how to drive more economically. Few automobile owners tinker under the hood anymore, not only because of the black boxes of electronics that reside there but also because of the tight space in which the engine rests. However, there are other ways to tinker with a vehicle, and sometimes a seemingly small design change in something that is all but ignored or taken for granted can make a measurable difference. One veteran truck driver invented a "fuel-saving mud flap" by making it screenlike, in contrast to the usual solid rubber sheet. Its inventor claims that by allowing as much as 75 percent of the air to pass through its many slitlike openings, the flap's "wind drag" is reduced, resulting in a fuel savings of 3 percent. During the 2008 election campaign, presidential candidate Barack Obama took some ribbing for suggesting that as a means of conserving gasoline drivers should be sure that their tires are properly inflated.[39]

Newspaper, magazine, and television stories about gas mileage became very popular, as did consumer reports on hybrid, all-electric, and other nontraditional vehicles. Many environmentally conscious drivers had bought hybrids years earlier, not only to get more miles per gallon but also to reduce the emissions they left behind on the highway, part of their car's "carbon footprint." Not a few hybrid purchasers were disappointed, however, finding that to get the very optimistic mileage advertised by the manufacturer they had to accelerate at a very carefully learned and monitored rate. Manufacturers and purchasers of all-electric cars chose not to mention the fact that it was often a fossil-burning power plant that produced the electricity that charged them.[40]

The fuel-cell-powered car is held out as the promise for a future of vehicles with benign emissions. The principle typically involves using hydrogen as a fuel to drive a chemical reaction that produces little else but electricity and water, which puffs out the tailpipe as harmless vapor or drips from it as clean liquid H_2O. Unfortunately, unbound hydrogen is not readily available in nature, and so vast quantities of it would have to be produced to feed into fuel cells. At present, the only proven and economical ways to generate considerable amounts of hydrogen are to extract it from fossil fuels or to burn those fuels to drive an electrolytic process. It would take the use of a renewable (and virtually free) resource, such as wind or solar energy, to accomplish the generation and use of hydrogen with a truly net reduction of fossil fuel use. Such drawbacks to using hydrogen as a fuel have prompted engineers to design fuel cells that operate on other gases, such as methanol, butane, and propane. However, until recently these flammable substances were defined as "hazardous materials" and so could not be carried on airplanes, precisely where early users of fuel cells would want them to power their laptops and other consumer electronic devices. With this regulation relaxed, it was expected that the perennial "technology of tomorrow," which for seeming decades has been only a year or two away, might finally move ahead. Panasonic has promised by 2012 a "fuel cell that can power a laptop for 20 hours on a cup of methanol." But not everyone can wait for improvements in powering consumer devices.[41]

One New Jersey resident, civil engineer Mike Strizki, employs solar panels mounted on his garage to provide the power for an electrolyzer that breaks down tap water into its constituent elements. The oxygen is vented to the atmosphere and the hydrogen is stored in old propane tanks, which, since hydrogen is so much lighter than air, weigh less when full than when empty. During the summer, the solar panels generate on average about ninety kilowatt-hours of electricity each day; this is enough not only to run the electrolyzer but also to provide all the power needed to run Strizki's house and to charge a bank of batteries that provides electricity when the sun is not shining. He uses hydrogen in fuel cells that power those of his vehicles and machines, including a lawn mower, that do not run on electricity.

Strizki boasts that he is energy self-sufficient and has no electric, oil, or gas bills to pay. Since water and sunshine are his raw materials, he can be enthusiastic in every type of weather: "If it's raining, it's fuel. If it's sunny, it's fuel. It's all fuel." But his self-sustaining "off-grid energy system" did not come as cheaply as the rain and sun. "Mr. Gadget," as he is known to friends, spent $100,000 of his own money, to which the New Jersey Board of Public Utilities added $400,000 in grants. There were also in-kind contributions from technology firms, but whatever was used in assembling the system was off-the-shelf technology. A full accounting should also include the brainpower and sweat equity he put into his grand suite of gadgets, but the dollar amounts for hardware alone are what make skeptics wonder if the world will operate on hydrogen anytime soon.[42]

The introduction of fuel-cell-powered cars on a large scale provides a classic situation of producing a product that does not fit neatly into the prepared context of an existing infrastructure. Since the vehicles run on hydrogen, it would be necessary for drivers of such cars to have a means of refilling their hydrogen tanks. But hydrogen filling stations are not nearly as common as gasoline stations, and they are not likely to be so until there are enough fuel-cell-powered cars on the road to patronize these stations and make them profitable. The situation might seem to present a standoff, with fuel-cell car manufacturers holding back production until refueling stations are in place and investors in hydrogen filling stations sitting back until there are enough cars on the road. In the end, manufacturers will likely go ahead with car production, assuming that the stations would follow as surely as gasoline stations and improved roads followed the introduction of Henry Ford's early internal-combustion-engine-powered cars.

The situation has proved not to be so simple as "if you build them, they will come." There would also be a matter of developing the technology to the point where the vehicles would have a sufficient range on a tank of hydrogen—and still be affordable. When the electric-gasoline hybrid Toyota Prius was introduced, prospective owners had to be willing to spend on the order of $3,000 to $5,000 more for it compared to an otherwise equivalent conventionally powered vehicle. Some committed environmentalists did pay the premium, accepting it as the cost of doing their small part for the good of the planet. There will also no doubt be a premium placed on early fuel-cell cars.

In the summer of 2008, a convoy of hydrogen-powered vehicles set out from Billerica, Massachusetts, which is about twenty miles northwest of downtown Boston and the location of the state's first hydrogen refueling station. This convoy, headed for California, did not face the inadequate roads that the U.S. Army did in 1919 on its famous sixty-two-day transcontinental journey to promote the Motor Transport Corps and demonstrate its defense mobility. However, like that military adventure, the hydrogen convoy had to be accompanied by a fuel truck. The difficulties that the army encountered remained in the mind of a young lieutenant colonel, Dwight Eisenhower, who, when he became president, was determined to institute an interstate highway system. The hydrogen convoy benefited from that network of roads, expecting to make the journey in thirteen days, including promotional stops, and its participants were exploiting the convenience of the interstates to pursue a dream of their own. Among the car manufacturers represented in the hydrogen convoy were BMW, Daimler, General Motors, Honda, Kia, Nissan, Toyota, and Volkswagen. They could not all be leaders of the convoy at the same time, but each no doubt wished to be a leader in the hydrogen economy they envisioned for the future.[43]

The presence of the hydrogen fuel truck in the convoy emphasized the fact that there were few refilling stations between the East and West Coasts. One logical way to introduce such facilities would be to add hydrogen pumps to those dispensing fossil-based fuels at existing service stations. However, to do this would mean that oil companies would have to make capital investments in new equipment that would in effect enable them to dispense a new fuel that would be in direct competition with their traditional product, gasoline, which is not likely to be obsolete soon. In California, free-standing hydrogen fuel stations have been established by the California Fuel Cell Partnership, a collaborative effort of automobile manufacturers, energy companies, fuel-cell technology companies, and government agencies. The thirty-six stations are clustered around

San Francisco Bay and Los Angeles. Across the entire country, as of mid-2008, there were on the order of one hundred hydrogen stations.[44]

In contrast, at the same time there were some 150,000 conventional gasoline stations, supported by an infrastructure of refineries, pipelines, and tankers that had developed over the century since the introduction of the internal combustion engine. A study conducted by General Motors concluded that an adequate hydrogen supply infrastructure might consist of twelve thousand hydrogen stations. If these were located in the largest one hundred U.S. cities, 70 percent of Americans would live within two miles of a station. Collectively, the system could support a million hydrogen-powered vehicles. Until such an infrastructure is established, drivers of these vehicles cannot range across the country at will the way drivers of gasoline-powered vehicles can.[45]

Existing hydrogen fueling stations were used by prototype fuel-cell-powered vehicles, which generally were being leased from the manufacturers. General Motors, which had a rented test fleet of Chevy Equinox sport utility vehicles on the road, was expected to be selling the model by 2010. Honda's FCX Clarity was being leased in Southern California. Toyota and Honda, whose prototype vehicles have been reported to cost the automakers as much $1 million apiece to manufacture, had a goal of bringing the production cost down to about $50,000 by 2015. Among the factors keeping the price up was the need to use platinum in the fuel cells. A 2008 government report projected that a best-case scenario had manufacturers selling about two million fuel-cell vehicles by 2020, but this would represent less than 1 percent of all vehicles on the road. A Ford engineer placed the date closer to 2030, due to the difficulty and expense of obtaining fuel. As a rule, considerable engineering is involved in taking something from the prototype or test stage to mass production, and fuel-cell vehicles cannot be expected to be an exception. The bottom line is that it remains to be seen whether fuel-cell-powered vehicles will result in a net plus for energy conservation.[46]

One immediate alternative to hydrogen is compressed natural gas, promoted as CNG and said to be the "cleanest of fossil fuels." The increased use of this domestic resource is a central component of the plan to reduce America's dependence on foreign oil promoted by the wealthy oilman T. Boone Pickens. Because there is concern among large users of natural gas—including electric utilities, which rely on it for about 20 percent of their generating capacity, and the chemical industry—Pickens also promotes the large-scale use of wind turbines to generate electricity, thereby freeing up natural gas supplies to power motor vehicles. His plan is not without his own self-interest, for Pickens is heavily involved in natural gas and wind energy businesses, but he emphasizes the energy-independence aspect of the plan.[47]

Natural gas has been advertised as "*the* bridge fuel to hydrogen" and has been promoted as a clean alternative fuel. By one estimate, burning natural gas can result in 23 percent less greenhouse gas emissions from a diesel engine and 30 percent less from a gasoline one. According to proponents of the Pickens Plan, the only thing that need be done to an internal combustion engine to convert it from burning gasoline to burning CNG is to change its fuel system. Since more than 70 percent of American homes are already connected to natural gas lines, the installation of a compressor in the garage would enable a CNG vehicle to be refueled there overnight. Similarly, service stations with a compressor and access to natural gas could easily be set up as natural gas refueling stations. Natural gas service stations continue to be located mainly in California and New York, where commuter and fleet vehicles are driven a predictable distance and return to a home base each day. This is important because CNG-powered cars like the Honda Civic GX, claimed variously to be the "greenest vehicle" and the "cleanest internal-combustion vehicle in the world," have a range of only about 250 miles. Within current technology, a greater range would require a larger fuel tank, which would take more car space to accommodate. Engineering is the art of compromise under constraint.[48]

The Civic GX has been available since 1998, but mostly to fleet owners and operators. Ford and General Motors, which ceased production of their natural gas vehicles before gasoline prices peaked, were considering getting back into the business. In late 2008, Toyota unveiled its Camry hybrid, which burns CNG rather than conventional gasoline. Outside the United States, natural gas has already been much more widely adopted as a fuel. As of 2008, there were more than seven million natural-gas-burning vehicles in operation worldwide. Among the countries having the most such vehicles were Argentina, Brazil, Pakistan, and Thailand.[49]

In the meantime, when gasoline prices escalated in the summer of 2008, many drivers of SUVs looked to trade them in for more fuel-efficient cars, but with fuel prices so high, there was not much market for SUVs, whether new or used, and so their trade-in value plummeted. An SUV owner had to take a considerable loss on his or her investment in buying a smaller vehicle, an amount that very often would have taken many years to make up in cumulative savings on gasoline. At about the same time that such calculations were beginning to be made in earnest, automobile manufacturers were introducing hybrid SUVs and pickup trucks, priced at a premium (often in excess of $50,000) for their environmental cachet, even though some were getting only twenty miles per gallon.[50]

Measuring fuel efficiency in miles per gallon itself has been challenged as nonintuitive when trying to compare different vehicles and gas-saving strategies. Thus, it may not be immediately obvious which of two options saves more gasoline: (1) trading in a compact car getting 34 miles per gallon for a hybrid that gets 54, or (2) replacing an SUV getting 18 miles per gallon with a sedan that gets 28. Using the alternative measure of gallons per 100 miles, the options can be restated as: (1) a compact using 2.9 gallons per 100 miles versus a hybrid using 1.9, or (2) an SUV using 5.6 gallons per 100 miles versus a sedan using 3.6. In this latter means of comparison, it is clear that the *savings* is twice as much by choosing the second option. Since it is the cumulative savings in outlay for gasoline over a period of time that should be compared with the capital outlay involved in making a switch, thinking in gallons per mile is much to be preferred. This is why fuel efficiency is quoted in liters per 100 kilometers in many European countries.[51]

Some American automobile manufacturers appear to have been focusing on the development of less numerate information for drivers. The 2010 Ford Fusion and Mercury Milan gasoline-electric hybrid midsize sedans were expected to have "Zen-like instrument panels" with electronic displays on which "images of green leaves multiply when drivers are motoring in fuel-efficient fashion." Evidently the auto company believed that "there is an appetite for this kind of thing, to make the vehicle more engaging and more entertaining while driving more ecologically and efficiently"; it wished "to create an emotional connection and 'wow' factor." Does this kind of thinking bode well for the future of the American automobile? Are Detroit engineers and their marketing partners focusing more on entertainment value than on fuel efficiency or safety?[52]

No matter how expressed or exploited, fuel efficiency generally correlates with the size of a vehicle, which in turn can correlate with cost. Engineers at the Indian auto manufacturer Tata Motors returned to basics to design the Nano, the world's least-expensive car. This "people's car" evoked the 1950s Volkswagen Beetle, down to its front trunk that can hold not much more than a computer case. The expected price of the car was estimated to be as low as $2,000, a figure to be achieved by such value-engineering design decisions as eliminating what are in fully developed countries generally expected features, like a radio, power steering, and air-conditioning; equipping the car with one rather than two windshield wipers; and using an engine that might be found on a fancy riding mower. Another factor in keeping the initial sales price low was the Indian government's acquisition of a thousand acres of farmland on which Tata could build its manufacturing plant. However, when political opponents claimed that the displaced farmers were not fairly compensated for their land and encouraged them to protest, construction was halted and the motor company had to regroup.

Fortunately, the manufacturer had a backup plan to move production to other of its plant facilities, even if it meant reducing the anticipated initial output from forty thousand to ten thousand cars per month. Regardless of where it would be produced, the Nano would be a far cry from a luxury car like a Lexus. The expected initial sales price of the tiny car was reported to be "less than the cost of the optional surround sound system and DVD player on the Lexus LX 570 sport utility vehicle." Furthermore, according to one report, the "downside is a car that would likely fail emission and safety standards on any Western road, and, perhaps, in India in a few years, when the country imposes tougher environmental standards." Nevertheless, the achievement was seen as an opportunity for India to "export a kind of 'Gandhian engineering,' combining irreverence for conventional ways of thinking with a frugality born of developing-world scarcity."[53]

Whether frugal or not, engineering is all about designing devices and systems that satisfy the constraints imposed by managers and regulators. The engineer's charge may be to design a car costing less than a certain amount of money or emitting less than a certain amount of carbon dioxide. Whichever may be the overarching goal, the way engineers will reach it is fundamentally the same—they will consider alternatives and choose those that move them closer to the goal. Often, the constraints are competing, as in the case of an engineer's being tasked with designing a car that is both cheaper and less polluting. In such a case, compromises have to be made; typically, either the price goal or the pollution goal or both will need to be relaxed. If there are environmental regulations that do not allow for meeting the pollution goal economically, then obviously the price has to rise. This is the reality of creating something new, and it is what engineers have always done and always will have to do.

10

Complex Systems

Among the most visible campaigners for environmental causes has been former vice president Al Gore. In speaking about the difficulties he has faced in communicating environmental threats to planet Earth, he has said, "One of the many complexities that complicate the task I've undertaken is complexity." Unfortunately, there is no entry for the word *complexity* in the index to Gore's book on global warming, *An Inconvenient Truth*, nor are there entries for *engineer* or *engineering*. He may consider communicating the inconvenient truth of what he calls a "planetary emergency" complicated by complexity, but so is the physical and mathematical complexity of responding to the emergency with realistic solutions. Indeed, according to some observers, "global warming's complexity and momentum have led to a try-everything approach by scientists," a group that might include, one would presume, engineers.[1]

Most scientists and engineers concentrate their efforts in a specialized field and so do not have much experience with systems as complex as the Earth's biosphere, which includes the surface of our planet and its atmosphere in which living organisms reside. Buckminster Fuller, who called himself "a 'comprehensive, anticipatory design scientist'—a 'comprehensivist,' for short," referred to our planet as Spaceship Earth, which better conveys the fact that we inhabit a fragile global ecosystem in which we are being transported across enormous distances over long periods of time. As the accidents with the space shuttles Challenger and Columbia demonstrated, something just a little out of order on one part of the vehicle can have profound effects on another part and on the safety of the whole system and mission.[2]

Our Spaceship Earth has two sources of energy to continue its journey: that which we carry along with us and that which we can harvest from incoming sunlight. Of the first, we have mainly thought about—when we have thought about energy at all—fossil fuels, those residua of eons past that have been pressure-cooked to our liking. We have burned coal, oil, and natural gas to heat compartments of our spaceship; to generate electricity to provide light and power for our small appliances and the larger machines of industry that we carry aboard our craft; and to drive the transportation devices that take us from place to place around our spherical ship of enormous girth.

We have also employed the forces of the wind and of gravity, that mysterious field that converges on the center of our vehicle, keeping it together as an aggregated whole. It is gravity that ultimately drives our great hydroelectric power plants connected to the perhaps tens of thousands of dams, some, like the Three Gorges Dam in the far eastern sector of our ship, of gargantuan proportions and unprecedented output. In addition to being renewable, this is considered clean energy, since its use does not emit noxious gases or produce contaminated by-products that cannot be easily jettisoned. But, as we have learned on our journey to essentially the same place in the solar system where we were one, two, and many years ago, we have to take the bad with the good.

Although some spaceship passengers may think hydropower is a perfectly clean energy source, others think it comes at too steep a price. In order to develop what engineers call a "hydraulic head," we have to impound a considerable depth of water; this is achieved by building a large dam, which itself takes an enormous amount of energy. The best location for a dam, especially a large one, is on a river that flows through a deep gorge or valley with a sufficiently large volume to keep the reservoir created behind the dam full and thus make the construction effort worthwhile. Such a location is often in a remote wilderness area of uncommon beauty.

Because dams create reservoirs whose stores of water drive the turbines that generate electricity, the footprint of a hydroelectric project can spread for hundreds of miles behind the concrete monolith that is dwarfed by its effect. Just as the hot exhaust trails of spaceships during launch consume anything combustible that they engulf, so the reservoirs that form behind dams inundate the natural canyons, gorges, and valleys that define their reach. When a potential reservoir covers land that is inhabited, building the dam destroys plant and animal life (whose rotting produces greenhouse gases like methane), displaces people, and submerges antiquities that embody lives past.

Few built things have a single purpose. Although they may be called hydroelectric dams, these structures may also be designed to provide flood control downriver, recreational lakes upriver, and water supplies for irrigation and for urban areas hundreds of miles away. Often, one or two of these other uses are at least as important as the primary one. This is the case with the O'Shaughnessy Dam, which after its completion in 1923 converted the Hetch Hetchy Valley, said to have been equal in natural beauty to nearby Yosemite Valley, into a reservoir to supply clean and reliable mountain water to the city of San Francisco, 250 miles away, and to generate electricity, which the city began to sell to the utility company Pacific Gas & Electric. The proposal to dam the valley brought about the first great environmental challenge of the young Sierra Club. Initially led by the environmentalist John Muir, the case went all the way to the U.S. Supreme Court, which ruled in favor of the dam, thus in effect siding with the city by the bay so far away.[3]

After it had been in place for almost half a century and had become well established as an essential part of the infrastructure of San Francisco, environmentalist descendants of Muir called for the removal of O'Shaughnessy Dam and the restoration of Hetch Hetchy Valley to its natural state. Whether ecological restoration is a realistic goal is a question for science; whether and how the massive concrete structure can be safely dismantled is a question for engineering. There have been increasing calls for the removal of other dams, some of which merely impound water and some of which once were the source of hydroelectric power for local industry or an area population.[4]

The first dam across the Kennebec River, at Augusta, Maine, was constructed in 1837. When Edwards Dam, as it was known in the late twentieth century, was up for relicensing, the Edwards Manufacturing Company and the city of Augusta must have believed that the

Federal Energy Regulatory Commission would grant their request to continue operating the 3.5-megawatt hydroelectric plant. FERC had viewed dams as permanent structures, and so the expectation was that they should remain in place. However, in 1997, FERC, perhaps being influenced by a coalition that had been formed to advocate removal of the Maine structure, came to believe that, like everything else, "dams have a finite life cycle." The commission not only denied a license renewal for Edwards Dam but also ordered it to be removed at the owner's expense. In reaching this unprecedented decision, the commission had considered three options: (1) approving as submitted the request for relicensing; (2) requiring the installation of a fish passage system, representing an investment of about $10 million; and (3) denying a renewal license and demanding the removal of the dam. Not only would installing a fishway be almost twice as costly as the last option, but removing the dam also would provide increased wetland habitat and more recreational waterway. It would benefit fishing, too, by opening up the spawning habitat of Atlantic salmon, Atlantic sturgeon, striped bass, rainbow smelt, and other migratory fish to historic upstream limits.[5]

With the handwriting on the concrete wall, the owners of the dam agreed to turn it over to the state. This absolved them of any deconstruction expense. In preparation for the removal of the dam in 1999, a dirt-and-gravel cofferdam was constructed to hold back the water so demolition work on Edwards could begin. Initially, a 75-foot section of the 24-foot-high, 917-foot-long main dam was removed, with the rest to be demolished after the reservoir drained. When water began to wash away loose gravel as a backhoe was being used to breach the cofferdam, the machine and its operator were almost lost. According to a fishing guide, after the breach there were "acres of dead fish backed up in coves in Merrymeeting Bay," in which the waters of the Kennebec mix with those of several other rivers on their way to the Atlantic. He also reported that a "dirty plume could be seen three miles out into the ocean." We live about thirty miles downriver from the dam and were watching the water on the day of the breach. Had we not known of the operations in Augusta, we would not have realized that anything remarkable had taken place. Just as we did before Edwards Dam had been removed, we regularly see enormous sturgeon jumping out of the water on their way upriver, safe from the eagles only because of their size.[6]

Edwards Dam was only one of about 75,000 dams across the country that are at least five feet high. While some extremists might like to see all such dams removed, more moderate environmentalists concern themselves only with those whose environmental and economic costs are not seen to be exceeded by their benefits. That would amount to only a small fraction of the total, according to the president of American Rivers, which advocates dam removal. In some cases, the environmentalists get well out in front of the issue. The Elk Creek Dam, on a tributary of Oregon's Rogue River, was never even finished before it was removed. Construction on the structure was halted by a federal judge in 1987, when it was about a third complete, in order to conduct an environmental review that considered what effects the dam would have on the ecology of the stream used for spawning by the threatened coho salmon and other species. After more than twenty years of debate over the dam's fate, the "concrete plug" was dynamited in 2008.[7]

Removing those dams, like Edwards, that have long been completed can take considerable study and preparation not only in terms of engineering but also in terms of science, finance, and politics. Four aging hydroelectric dams on the Klamath River in Oregon and California had long come under attack as the cause of reduced stocks of fish. And not unlike the case with Edwards, the dams' owner found itself facing the considerable expense of building new structures and systems to restore access to the salmon in order to receive a license renewal. Rather than incur this uncertain expense, the utility PacifiCorp entered into a tentative agreement with the states and the federal government to remove the dams by the year 2020. The removal would be financed in part by a surcharge on the power generated by the dams and supplied to some 600,000 regional customers. A final agreement and plan would come only after "several years of scientific analysis of how removing the dams would affect water quality and fish habitat . . . to ensure that the environmental impact of the release of sediment would not be worse than leaving the dams in place." Since the time frame encompassing the provisional agreement, the studies, and a final decision could span as many as three or four administrations in Washington, there was some concern about how faithfully plans might be carried out. A spokesman for PacifiCorp was optimistic that there would be continuity of purpose because the process was driven by "science—unbiased, unadulterated science."[8]

Historically, less attention was paid to environmental effects than is presently the case. The construction (or deconstruction) of a dam, even a relatively modest one like Edwards, is no trivial design problem. In order to build a dam that will successfully hold back a portion of a river, the location must be chosen with an eye toward the suitability of the river bottom to provide a sound foundation. For larger dams, consideration must be given to the stability of canyon or valley walls or riverbanks to serve as sound abutments, and the nature of the topography upriver to accommodate the amount of water that will be stored. To these strictly technical matters must now be added considerations of how the dam might impede the movement of fish and other wildlife up and down the river. The evaluation of a prospective site involves a careful survey of the area and its ecology.

If a single dam cannot be looked at in isolation from its natural, social, political, and ethical implications, then how much more so must systems thinking be necessary to comprehend the interactions among the various components of Spaceship Earth. The complex interrelationships among the oceans, atmosphere, winds, solar rays, glaciers, ice fields, flora and fauna, and every other meteorological, geological, and ecological aspect of our planet have become increasingly clear. It is as if our hurtling spaceship were the Apollo 13 mission, with malfunctioning components that need to be fixed by the people on board using only things and tools aboard.

Global warming, climate change, greenhouse gases, carbon footprints, the ozone layer: these phrases have been prominent in headlines, some for at least the last decade or two. Are they distinct causes, effects, beginnings, or ends? Or are they all related? Are they natural phenomena, or are they caused by human activities like burning fossil fuels? Should we be concerned about them and change our ways? Or should we just shrug each of them off as another Chicken Little scare? How much do we know about all this? These are questions for science, both natural and political.

According to the booklet *Understanding and Responding to Climate Change*, which contains highlights from relevant National Academies reports, "the phrase 'climate change' has been growing in preferred use to 'global warming' because it helps convey that there are changes in addition to rising temperatures." Regardless of one's political persuasion, there can be little doubt that politics is but another factor that must be taken into account in trying to sort out scientific facts from scientific fuss. The prominent note in the presumably neutral National Academies booklet distinguishing between global warming and climate change can be interpreted also as a recognition of the fact that "global warming" had become a politically charged phrase that was better avoided than used, lest it produce knee-jerk reactions among the committed and the skeptical alike.[9]

There are too many examples of clear objectivity and honest admissions of insufficient knowledge in National Academies reports to claim that they are blatantly biased, but there are subtle indications in the written record that some things are assumed to be true that perhaps should only be stated as possibilities or probabilities. Consider the opening paragraph of the *Understanding and Responding* booklet, the very title of which implies that climate change, that is, "global warning," is a fact to be understood and responded to:

> There is a growing concern about global warming and the impact it will have on people and the ecosystems on which they depend. Temperatures have already risen 1.4°F since the start of the 20th century—with much of this warming occurring in just the last 30 years—and temperatures will likely rise at least another 2°F, and possibly more than 11°F, over the next 100 years. This warming will cause significant changes in sea level, ecosystems, and ice cover, among other impacts. In the Arctic, where temperatures have increased almost twice as much as the global average, the landscape and ecosystems are already changing rapidly.[10]

The paragraph begins with an arguably neutral "growing concern," progresses through some probabilistic conditions that "likely" might and "possibly" could occur, and then moves into "this warming will cause" scenarios. In other words, it progresses from hypothesis to fact in the course of a hundred or so words. What is not stated is: Who has this "growing concern"? How likely is it that temperatures will rise to the extent stated? What constitutes a "significant change"? Scientists as well as lay readers should want to know the answers to such questions, and the booklet should anticipate those questions. If engineers are to be charged with seeking solutions to the potential problems associated with climate change, then policy makers should have much more information on which to base their judgment. If tools such as quantitative risk assessment are to be employed in the decision-making process, then the best possible data should be available. This often entails reading deep into the technical literature and separating the wheat from the chaff.

On the plus side, the booklet does acknowledge that, "in some parts of the world, global warming could bring positive effects such as longer growing seasons and milder winters." But, the booklet goes on, "it is likely to bring harmful effects to a much higher percentage of the world's people," such as those who live in coastal regions, where rising sea levels can be expected to bring increased flooding. Indeed, "there will be positive and negative impacts of climate change, even within a single region." Thus, warmer temperatures could prove to be "benefiting those farmers who can adapt to the new conditions but potentially harming native plant and animal species. In general, the larger and faster the changes in climate might be, the more difficult it would be for human and natural systems to adapt."[11]

In fact, there are counterbalancing pluses and minuses involved not only in the anticipated end effects of climate change, but also in the very natural processes by which those effects could be reached. Consider the nature of the ozone hole. Scientists have long believed that the ozone layer of the upper atmosphere plays a positive role in shielding the Earth's surface from solar ultraviolet rays, whose effects can be harmful to humans. Decades ago, it was discovered that each summer a hole in the ozone layer opens up, a phenomenon that scientists attributed to chemical pollution due to the use of chlorofluorocarbons, such as the Freon then used in refrigerators and air conditioners. After 1996, when CFCs were banned by an international treaty, the annual ozone hole began to shrink, and scientists predicted that in time it might stop forming altogether. This was considered a positive development.

However, some atmospheric scientists imagined a negative consequence of a disappearing ozone hole, one that might adversely affect climate change. Based on results from their models, they reasoned that global warming pushed toward the South Pole westerly winds that blow around Antarctica, over which the ozone hole forms. Since these winds affect "everything" from the extent of ice floes to the location of deserts in the Southern Hemisphere, they could also increase the rate of global warming. With a permanently closed ozone hole, wind patterns would be greatly altered, moving toward the equator, and thereby bringing more warm air to Antarctica, whose ice shelves would melt faster. Also, without an ozone hole through which to escape, greenhouse gases would accumulate below the ozone layer and trap additional heat. These dire predictions were based on computer models that did not include the effects of the oceans, which the scientists admitted could have a profound influence on the outcome. In other words, these, like many predictions, were a work in progress.[12]

Scientific reports relating to the Earth's climate not infrequently appear to be downright contradictory. The following two headlines appeared in *The New York Times* almost exactly one year apart:

GLOBE GROWS DARKER AS SUNSHINE
DIMINISHES 10% TO 37%

EARTH HAS BECOME BRIGHTER;
BUT NO ONE IS CERTAIN WHY

Could both be true? Did so much change in a single year? The first story related the belief that pollution particles reflect light back into space and lead to thicker and darker clouds, which block out light. The range of percentages reflected the variation of the effect from place to place on the planet. The second story reported that, after decades of trending to "global dimming," the surface of the Earth was becoming brighter. Was this a temporary thing, or was it a harbinger of a new decades-long trend? Elsewhere, we could read that global dimming may be masking effects of global warming, and so cleaning up the atmosphere might allow the Earth's temperature to rise at greater rates. Headlines obviously tell only part of the story.[13]

There are also opposing views on the question of whether the sun, which itself is slowly growing brighter as well as larger, will eventually consume the Earth. Over the course of almost a century, it has been reported, a "bevy of scientists have reached oscillating conclusions. In some scenarios, our planet escapes vaporization; in the latest analyses, however, it does not." As with the prediction of the path of a hurricane, the divergent opinions result from the use of different computer models with different underlying assumptions about the complex interaction among the bodies of the solar system. As the sun ages and loses mass, its gravitational pull on the Earth is lessened, which can allow our planet to travel in an orbit farther from the sun and hence beyond its damaging effects when it reaches red giant size. But, as the sun grows, its more ethereal surface can induce drag on the Earth, thereby slowing it down and consequently pulling it toward the sun. When the gravitational interaction of the rest of the planets of the solar system is taken into account, the prediction of the orbit of the Earth becomes much more complicated. With complexity comes increasingly difficult mathematical problems to solve—even with the world's largest computers—which leads different scientists to come up with different approaches to simplify the calculations, which in turn leads to divergent predictions coming out of what are black boxes to the uninitiated and even to some of the initiated. What's a person or an engineer, or a scientist for that matter, to think?[14]

The phenomenon of each new scientific finding going contrary to the last one has been described as a "windshield wiper effect." This frustrating back-and-forth movement in policy and debate in science is frequently seen in the health sector, with one study finding that a substance is good for us and another that it is bad. Even something seemingly as straightforward as testing middle-aged males for a prostate-specific antigen has urologists taking sides. Men diagnosed with prostate cancer—and their doctors—have found themselves thoroughly confused about what course of action to take because of so many conflicting reports. Outside the health and medical fields, the windshield wiper effect has also been observed by ordinary citizens trying to follow the debates over global warming engaged in by scientists.[15]

Predictions of the frequency of occurrence, the intensity, and the path of hurricanes obviously provide essential information for calculating risk, making policy decisions to guide the design and construction of defenses, and ordering evacuations. However, the weather and the modeling of it are not sufficiently well understood to give definitive answers. Will hurricanes occur less or more frequently as a result of global warming? The answer depends on which group of scientists is asked. According to one report of research findings from the National Oceanic and Atmospheric Administration, because of climate change the frequency of occurrence of hurricanes and tropical storms will diminish by century's end. But even though there may not be as many hurricanes over the next century, those that do occur can be expected to be slightly more intense. These findings are in direct contradiction with those of other researchers, who predict that a warmer world means more hurricanes.[16]

Where uncertainties about influences on climate change remain, some scientists and policy makers understandably call for more theoretical, experimental, data-collecting, and modeling efforts to add to our store of knowledge and tools, which might help them come closer to unambiguous conclusions about the state and future of Earth. In the meantime, the 2007 report of the Intergovernmental Panel on Climate Change, which the United Nations oversees, attributed a "greater than 90 percent certainty to the conclusion that human activities were driving the recent global warming trend." Furthermore, the panel was reported to have used "expert judgment" to place the confidence level at greater than 50 percent "that there had already been an increase in intense hurricane activity, partly caused by global warming."[17]

To skeptics, calling for more research—and more money to fund that research—can seem self-serving. To scientists, it is precisely the skepticism of those who do not embrace the idea of climate change that requires that more money be spent to make the case for adverse climate change more convincingly—or to disprove it. To be fair, some scientists have said that the evidence is already so strong that measures to curb carbon dioxide emissions should be undertaken even if the jury is still out. In mid-2008, the atmospheric concentration of carbon dioxide was said to stand at its highest level in 650,000 years, or at about 385 parts per million, which is approximately 38 percent higher than the pre–Industrial Revolution level of 280 parts per million. A full-page ad in *The New York Times* warned that this concentration was already too high, and that steps should be taken on an international scale to reduce the level back to what it was twenty years earlier, when James Hansen, of NASA's Goddard Institute for Space Studies, told Congress that human activity was responsible for generating greenhouse gases that were leading to a warming of the globe. At the time of Hansen's testimony, the CO_2 concentration was 350 ppm.[18]

While questions remain unanswered, there does appear to be sufficient evidence to be concerned about climate change and to take certain measures to mitigate it—including working to reduce the quantity of greenhouse gases in Earth's atmosphere. It is here that scientists should either hand the problem over to engineers or engage not only in science relevant to climate change but also in engineering means to control it. Among the engineering problems that need to be addressed are reducing undesirable vehicle and power plant emissions and developing cleaner fuels or power plants that can use innovative fuels. These are natural problems for scientists and engineers to join forces in tackling, perhaps engaging in global research and development programs whose goal is not new industrial products and processes per se but true professional cooperation for the common good. For example, international interdisciplinary teams would be appropriate for tackling problems relating to the rise (or fall) of ocean levels, there naturally being interested parties on all bordering shores. This kind of R&D might logically be funded on an international basis, but designing an entity to oversee such a program might be as difficult a political-engineering problem as the underlying challenge is a global-engineering one.

Sometimes seemingly good intentions for limiting pollution locally can have the opposite effect globally. Whenever old automobiles, trucks, boilers, and other polluting equipment are replaced in the United States by more energy-efficient and less air-polluting newer models, it would appear to be a net plus for the global environment. But this may in fact be true only locally, regionally, or nationally. The old machinery is typically sent to be reused as-is in developing countries, where it continues to belch out greenhouse gases. The effect, of course, is to pollute the air in another geographical area and to exacerbate the problem globally.

A particularly striking example relates to an old coal-fired plant that once powered a paper mill in Turners Falls, Massachusetts. The disused plant, which cost around $44 million to construct in the late 1980s, was disassembled and its 2,600 tons of components shipped to Villa Nueva, Guatemala, where it was reassembled to drive a textile mill that had been using an even dirtier-burning oil-fired plant. The next use of that plant was unknown but was very likely to involve reuse for some other purpose in that country. The cost of acquiring, dismantling, shipping, and rebuilding the replacement plant was about $22 million, and this was considered a wise investment, even though a greater amount of carbon dioxide is typically released by burning coal rather than oil. The international business in retired plants, vehicles, and heavy equipment has been estimated to be about $150 billion annually, but the overall cost to the environment is more difficult to ascertain. One thing is for sure, however, and that is that taking action to reduce pollution in the United States does not necessarily equate with reducing the emissions of greenhouse gases worldwide.[19]

The National Academies have considered hedging the bet over whether or not there will truly be catastrophic climate change. The approaches fall into the category of "no-regrets" strategies:

> Another way to prepare for climate change is to develop practical strategies for reducing the overall vulnerability of economic and ecological systems to weather and climate variations. Some of these are "no-regrets" strategies that will provide benefits regardless of whether a significant climate change ultimately occurs in a region. No-regrets measures could include improving climate forecasting based on decision-maker needs; slow-

ing biodiversity loss; improving water, land, and air quality; and making our health care enterprise, financial markets, and energy and transportation systems more resilient to major disruptions.[20]

Such approaches would, of course, rely heavily on engineering problem solving. But the belief that there are some kinds of technological fixes that can reverse the course of events has produced "massive schemes to reengineer the planet." Many of these are actually unrealistic from an engineering point of view.

One physicist has offered a bold solution to the global warming problem. He maintains that locating sufficient numbers of carbon filters around the world could "reel the world's atmosphere back toward the 18th century, like a climatic time machine." Based on the performance of a prototype he constructed and tested, it would take more than 65 million filters the size of boxcars and cost trillions of dollars annually to remove and store (or sequester) the necessary billions of tons of pollutants. It would be an enormous challenge to find sufficient amounts of clean energy to power the "orchards of filters" without introducing new sources of pollution. The scale of the effort would require orchards covering altogether an area the size of Arizona and present to the world an annual "vacuuming bill" on the order of $5.6 trillion. Imagine the response if an engineer had proposed such a solution.[21]

Another scientist, a geophysicist, has proposed using a "synthetic tree" that takes CO_2 out of the atmosphere a thousand times more efficiently than a comparably sized natural tree. Ideally, a stand of artificial trees would blend into the landscape the way some treelike cell phone towers almost do, but the proposed fake tree design resembles "a goal post with Venetian blinds" or a "massive flyswatter." According to the argument, the efficiency of the fake trees would greatly reduce the estimated 2.5 billion acres of real trees that would be needed to counter the world's carbon emissions. That number of natural trees would cover more than the entire area of the United States. Global solutions to global problems can involve out-of-this-world thinking resulting in astronomical numbers.[22]

Other proposals have included blocking sunlight so that it cannot warm the planet so effectively. While this might be done just by allowing pollution to spread, it might also be done "by constructing artificial volcanoes to blast sulfur particles into the atmosphere or by launching millions of tiny satellites into space and arranging them into a giant mirror." Another proposal for cutting down on how much sun reaches our planet calls for installing a physical sun-shade in space at a location where the sun and the Earth exert equal gravitational pull. This location, known as a Lagrangian point, is about a million miles away from Earth and changes as the Earth orbits the sun. So, the sunscreen would consist of literally trillions of autonomous spacecraft, launched over a period of some thirty years, keeping themselves in the proper position by means of com-puterized navigation systems. The total weight of the sunscreen would be about seventy thousand times that of the International Space Station. Among the arguments made against a sunshade idea is that it "would waste solar power." It is difficult to tell whether such proposals are meant to be taken seriously. They certainly are not the kinds of solutions engineers seek.[23]

Another proposal, put forth by the California-based firm Plank-tos, which "works to develop science and technology to address global warming and the declining productivity of the world's oceans," would "fertilize part of the South Atlantic with iron, in hopes of pro-ducing carbon-absorbing plankton blooms that the company could market as carbon offsets." However, a recent study was reported to have found that complex and incompletely understood microbial interactions might result in less carbon absorption than expected. While promising schemes may work logically and scientifically, they can be questionable from an engineering point of view, which must consider the economics and system dynamics of a design—and they can be dangerous, too, because the adverse effects of large-scale intrusions into natural processes cannot easily be known. In fact, according to some critics, this kind of "geoengineering," as such large-scale efforts are collectively known, "would inevitably produce environmental effects impossible to predict and impossible to undo."[24]

Several decades ago, environmental scientists first proposed "spreading very small reflective particles" upon some five million square miles of ocean in order to reflect more sunlight back into space. A "physicist and energy expert" termed this scheme "a wacky geoengineering solution" to the problem of global warming. The particles would likely pollute the seas and wash up onto beaches. Other potential schemes have looked to land-based reflection technology. According to one calculation, "if the estimated 360,000 square miles (less than 1 percent of the world's land surface) covered by urban rooftops and pavement were a white or light color, enough sunlight would be reflected back into space to delay climate change by about 11 years." Such a change could be "a one-time carbon-offset equivalent to preventing 44 billion tons of CO_2 from entering the atmosphere." That is said to be equivalent to keeping all of the world's automobiles in their garages for more than a decade. Of course, it is unlikely that such global cooperation could be achieved easily, but there are alternative geoengineering schemes. Similar Earth-benefiting effects could be achieved by "covering the Sahara with enormous sheets of white plastic, for instance, or painting the Black Hills of South Dakota white."[25]

Although engineers are used to dealing with large numbers, the numbers involved in global problem solving can be mind-boggling and their implications mind-numbing. In addition to geoengineering schemes, there are questions about emerging technologies like nanotechnology and robotics. According to William A. Wulf, former president of the National Academy of Engineering, "the complexity of newly engineered systems coupled with their potential impact on lives, the environment, etc. raise a set of ethical issues that engineers had not been thinking about." But, according to Andrew Light, director of the Center for Global Ethics at George Mason University, "it would be unethical not to embark on the work needed to engineer possible remedies" to global problems. Engineers are not averse to ethical considerations, but they also crave the elegant technical solution that makes both environmental and economic sense.[26]

Embarking on solutions is no trivial matter. According to Wulf's successor, Charles M. Vest, "engineering is about systems"; he told the Global Summit on the Future of Mechanical Engineering that "the frontiers of engineering today are in tiny systems on the one hand, and in macro systems on the other." In such fields as bioengineering, information technology, and nanotechnology, where very small scale is commonplace, Vest predicted that the distinction between scientist and engineer, who typically work closely together on problems, would virtually disappear. In dealing with macro problems relating to energy, environment, health care, manufacturing, communications, or logistics, he believes that there will need to be contributions from disciplines beyond engineering, including the social sciences, business management, and the humanities.[27]

In other words, the solution to problems involving complex systems can be expected to require the involvement of complex systems of people and approaches. Most, but not necessarily all, of those involved in seeking solutions will need to be scientists and engineers. Very big problems, such as those involving global climate change, are likely to engage large numbers of people, both technical and nontechnical, coming from a wide variety of fields and cultures and working in a number of interconnected but distinct geographical locations, in different time zones. This kind of global problem-solving effort has been used to design sophisticated airplanes like the Boeing 777 and can be expected to characterize project teams assembled, albeit largely virtually, to tackle macro-environmental problems. The management of such diverse and diffuse teams may best be handled by people with international experience, interdisciplinary knowledge, and intercultural sensitivities. Some managers may be scientists and engineers, but no one should expect such complex teams to function without a certain amount of friction.

11

Two Cultures

Half a century ago, the scientist and novelist Charles Percy Snow delivered a lecture at the University of Cambridge in which he described a problematic situation that he termed "the two cultures." According to C. P. Snow, as he came to be most commonly known, it was the circumstances of his involvement in both the physics and the writing communities, mostly in Britain, that gave him an unusually diverse perspective on intellectual life at mid-century. Although he noted that members of the two groups that he moved among had similar social origins, possessed comparable intelligence, and earned about the same amount of money, they barely communicated with each other. Snow observed that their "intellectual, moral and psychological climate had so little in common" that they may as well have come from different parts of the world. He feared that "the intellectual life of the whole of western society [was] increasingly being split into two polar groups"—epitomized by physical scientists and "literary intellectuals."[1]

Snow, no doubt using hyperbole, lamented the fact that those in the literary camp had little knowledge of or respect for science, while scientists had at best a passing acquaintance with and regard for books outside their field. The literati, who expressed "incredulity at the illiteracy of scientists," were guilty of behaving "as though the natural order didn't exist," Snow charged. And he ridiculed scientists who saw books only as "tools," rather than as sources of insight into the nature of humanity. In one memorable passage, he likened possessing a familiarity with the Second Law of Thermodynamics to having read Shakespeare. Snow further equated an understanding of such physical concepts as mass and acceleration with being able to read at all. He acknowledged that the dining tradition at British college high tables was to mix fellows from both the scientific and non-scientific disciplines, but he decried the fact that there was seldom significant intellectual intercourse between the two groups.[2]

Where engineers and engineering fit into Snow's dichotomy was ambiguous, and at times in his lecture he might have been accused of having had some ignorance of them. In his position of being in charge of wartime recruitment for Britain's scientific research efforts, Snow had been involved in interviewing a large number of "working scientists" and "professional engineers or applied scientists," which in this context he seemed to either equate or group together. On another occasion, Snow appeared to classify the Soviet Union's achievement of launching the first artificial satellite, Sputnik, as "one of the most astonishing experiments in the whole history of science," albeit "a feat of organization and a triumphant use of existing knowledge"—in other words, a feat of engineering. But engineering is much more than that. While sharing characteristics of the two cultures that Snow concentrated upon, engineering is also a culture unto itself and thus separate from each of them.[3]

We speak of the sciences and the humanities as if they were distinct collections of things—like the solar system or a biological genus—and we use these terms to imply a certain grandness of existence and superiority of position about which and within which can be made discoveries and pronouncements. But we do not use the collective noun and speak of "the engineerings," even though there are many. In fact, I don't recall ever before having seen the word *engineering* made plural. (Thus, we have a National Academy of Sciences, but a National Academy of Engineering.) *Engineering* is, in fact, always a singular noun in the form of a gerund, which my dictionary defines as "a verbal noun in Latin that expresses generalized or uncompleted action" and one that serves a similar purpose in English. Engineering is not only about creatively applying existing knowledge to ever-new things; it is also about ongoing process, even in the absence of existing knowledge. *Engineering* is an active noun, like *inventing* and *designing* and *making* and *building*, processes that are always continuing. Even when something is invented, made, or built, it can be improved upon and so again invented, made, and built better.

Perhaps even more so than science, engineering is akin to writing or painting in that it is a creative endeavor that begins in the mind's eye and proceeds into new frontiers of thought and action, where it does not *find* so much as *make* new things. Just as the poet may start with an empty sheet of paper and the painter with a blank canvas, so the engineer today (and perhaps also the writer and artist) begins with a computer screen devoid but for a maddening cursor blinking away the seconds of indecision. Until the outlines of a design are set down, however tentatively, there can be no appeal to science or to critical analysis to judge or test the design. Scientific, rhetorical, or aesthetic principles may be called upon to inspire, refine, and complete a design, but creative things do not come of applying the principles alone. Without the sketch of the thing or a diagram of a system or process in which they are to be exploited, scientific facts and laws are of little use to engineers. Science may be the theater, but engineering is the action on the stage.

Imagine wanting to build a bridge across a river. Clearly, Galileo's "two new sciences" are supremely relevant. The bridge must be able to withstand the internal stresses on its fabric resulting from the incessant pull of gravity and the intermittent forces on its roadway of traffic, snow, wind, and earthquakes. But knowing this does not alone produce a bridge. No matter how complete our knowledge of mechanics, without a geometric arrangement of the parts of the structure we have nothing to which to apply scientific knowledge; we have no boundaries on which to impose mathematical conditions. It is the geometry and method that the engineer supplies, usually first in a word picture or a rough sketch. Thus, a cantilever truss or a suspension bridge might be proposed, and only then can the tentative geometry be tested against the laws of mechanics. The process is typically iterative: as the design proceeds, revisions in one part prompt changes in another. When the process ends depends not on the laws of science but on the judgment of the engineer.

The French sociologist of science and technology Bruno Latour has likened a technological project generally to a work of fiction and has referred to engineers as "unsung writers." He recognizes that engineers write the narrative for the future, whether it be for better or for worse. Indeed, according to Latour, engineers "who dream up unheard-of systems always go further . . . than the best-woven plots" of fiction writers. This is entirely consistent—if not to be expected—from a discipline that von Kármán described as creating things that never were.[4]

For example, designing a bridge might be likened to writing a sonnet. Each has a beginning and an end, which must be connected with a sound structure. Common bridges and so-so sonnets can be made by copying or mimicking existing ones, with some small modifications of details here and there, but these are not the creations that earn the form its reputation or cause our spirits to soar. Masterpieces come from a new treatment of an old genre, from a fresh shaping of a familiar form. The form of the modern suspension bridge— consisting of a deck suspended from metal cables slung over towers and restrained by anchorages—existed more than half a century before John Roebling proposed his Brooklyn Bridge, but the fresh proportions of his Gothic-arched masonry towers, his steel cables and diagonal stays, and his pedestrian walkway centered above dual roadways produced the structure that remains a singular achievement of bridge engineering. And bridges are not the only engineered structures that evoke or are akin to poetry. A naval architect has observed that ship design "differs from the creation of poetry only in its numerate content." While some poetry reeks of numeracy, Shakespeare's sonnets, even though all containing fourteen lines of iambic pentameter fitted to a prescribed rhyme scheme, are as different from each other as one remarkable bridge or ship design is from another.[5]

The analogy between bridges and poems, like all analogies, can be carried only so far. Engineering structures must conform not only to their type but also to the laws of the physical world, which are unforgiving critics that cannot be refuted or ignored. A single undersized component precipitated the collapse of the Quebec Bridge in 1907, and ignorance of the aerodynamic effects of the wind brought down the Tacoma Narrows Bridge in 1940. Engineers cannot be granted poetic license, and they seize it at their own risk. Similarly, as was the case with Snow, otherwise masterful lecturers can invite scathing criticism with the slightest misstatement of the most incidental detail.

As was traditional with the Rede Lecture that Snow delivered in 1959 at Cambridge, the text of his "The Two Cultures and the Scientific Revolution"—to give its full title—was published as a pamphlet the day after it was delivered, and it was also excerpted in the magazine *Encounter*. There was some early feedback from readers, but Sir Charles (he had been knighted in 1957) was not prepared for the onslaught of commentary that eventually came. (An especially scathing and personal attack was delivered by the English literary critic F. R. Leavis, who termed Snow "portentously ignorant" and stated that his lecture "exhibits an utter lack of intellectual distinction and an embarrassing vulgarity of style.") By way of response to what he perceived to be misreadings, misinterpretations, and misplaced criticism, in 1963 Snow took a "second look" at his essay. This is appended to a "second edition" of the lecture, which is the text that I have used.[6]

At one point in his second look, Snow rephrases, as "quietly" as he can, the essence of his thesis:

> In our society (that is, advanced western society) we have lost even the pretence of a common culture. Persons educated with the greatest intensity we know can no longer communicate with each other on the plane of their major intellectual concern. This is serious for our creative, intellectual and, above all, our moral life. It is leading us to interpret the past wrongly, to misjudge the present and to deny our hopes of the future. It is making it difficult or impossible for us to take good action.[7]

Snow justified his use of the term *culture* in what was at the time an unusual sense for a word used mainly by anthropologists, though today we readily speak of academic, scientific, and literary cultures. Snow also defended his choice to divide the world into two, rather than many, intellectual cultures. But, mostly, he spent his time emphasizing and elaborating on his real intentions in the lecture, which included deploring the lack of appreciation that literary intellectuals had for the Industrial Revolution and to raise consciousness about the disparity between the rich and the poor in the wake of that revolution. (Snow admitted to almost titling his lecture "The Rich and the Poor.") At the root of both problems he saw the nature of the educational system, especially that in his own country, which promoted specialization. Snow felt that classical education was not preparing future opinion makers and political leaders to recognize and act on the implications of advancing technology, which he believed also eluded many scientists.[8]

Snow further attempted to clarify his view of science and technology. Though he considered it "permissible" for the purposes of the lecture to "lump pure and applied scientists into the same scientific culture," he seemed to have done so reluctantly. He acknowledged that "most pure scientists have themselves been devastatingly ignorant of productive industry," and remained so. Indeed, he asserted that "pure scientists and engineers often totally misunderstand each other," with the former being "dimwitted" about the latter, not recognizing that many engineering problems are "as intellectually exacting as pure problems, and that many of the solutions were as satisfying and beautiful."[9]

Snow admitted to at one time having tried to draw "a clear line between pure science and technology," but it was an effort he abandoned after seeing technologists at work. He found an engineering activity such as designing an aircraft to be an experience aesthetically, intellectually, and morally similar to setting up a particle physics experiment, and he found the "complex dialectic between pure and applied science" to be among "the deepest problems in scientific history." What Snow seems to have ignored, however, was that experimental physics relies greatly on technology, as has been frequently commented upon and discussed here. In fact, many advances in pure science would not have been possible without the technological development of apparatus and instruments that enabled observation and measurement. Though Snow declared that "the scientific process has two motives: one is to understand the natural world, the other is to control it," he may not have acknowledged sufficiently clearly the primacy of the latter motive—which is that of engineering—for his ultimate concern.[10]

For all the thought about the similarities and differences between intellectual endeavors that they evoked, C. P. Snow's "two cultures" lecture and essay ultimately were not about the rift between the sciences and the humanities so much as about the gulf between the technological and the human. In a sweeping accusation, Snow asserted that Western nonscientific intellectuals had "never tried, wanted, or been able to understand the industrial revolution," and that literary intellectuals were "natural Luddites." As a result, according to Snow, they failed to appreciate its implications for the future of the world and their moral responsibility in it. And though they dominated the ruling class and shaped opinion, they failed to foresee the changing responsibilities of the educational system. Those societies that experienced and benefited greatly from the Industrial Revolution did not appreciate all of its implications, while poor countries fully embraced industrialization as the way to improve their lot. In the latter, the study of engineering was not only encouraged but also respected.[11]

A great deal has happened in the half century since Snow lectured on the two cultures. His prediction of the industrialization of China has come to pass, and other then-poor countries are emerging as formidable players in the industrialized world. As Snow appreciated, much of this has to do with education and changing perceptions of industrialization. My own experience as an engineer, an academic, and a writer has given me a perspective that, while certainly not nearly so privileged nor so well placed as that of Sir Charles, has enabled me to see some changes from a special vantage point. Many of them would not have surprised him, but at least some undoubtedly would have.

As any engineering faculty member knows, many of his or her colleagues and their students have come from many different countries and cultures. Engineering is inherently an international and global profession, and experienced engineers have long taken new ideas from technologically richer to technologically poorer countries. Late-eighteenth-century and early Victorian engineers, having built canals and railroads in Britain, traveled to build them also in South America, Scandinavia, the Middle East, and elsewhere. Engineers who built railroads and dams across America took their know-how to Japan and China in the late nineteenth and early twentieth centuries. Engineers with experience continue to travel around the world, transferring technology from one culture to another. Until recently, the tallest building on the planet was in Malaysia, designed and constructed with the essential help of American structural engineers. As a by-product, Malaysians gained experience in working with high-strength concrete. Though Americans were involved when it was still only an idea, China virtually single-handedly built the world's largest dam, thus gaining further experience in a technology that China is now in a position to export.

Modern engineering problems can be worldwide endeavors. Design work on the Boeing 777 proceeded twenty-four hours a day, as engineers seamlessly passed, like a baton in a relay race, digital data and electronic plans from one time zone to another, picking up revisions from around the world on the next lap, the next working day. Who has not had a computer problem and talked on a help line with someone in India or Ireland? Technology has developed into a world culture, and its spread is not likely to be stemmed. Engineering education in America and other industrialized countries has long served a large number of international students, especially at the graduate level. Now, undergraduate classes also are being populated with increasing numbers of students from abroad. A natural corollary to this is that engineering faculties, which have long been internationalized, are becoming even more so.[12]

My experience as an academic has gradually extended beyond the classroom and even beyond the engineering school. When I first joined Duke University in 1980, the two-cultures problem was acute there. The humanities were prominent and privileged, at least from the perspective of a relatively small engineering faculty, and there was very little discourse across the divide, which was a literal one: humanities faculty occupied the signature stone buildings on the high ground of the campus; engineering and science faculty were housed down the hill, in redbrick structures across Science Drive, which might just as well have been called Science Divide. Crossing the street and making the trek up the hill took a deliberate act, and one that was seldom rewarded with intellectual discourse. Needless to say, few humanists came down the hill, even though it was an easier walk. In the faculty dining room, members of the individual cultures by and large ate and conversed at separate round tables, their backs arched like the tops of covered wagons, circled against attack, talking among themselves. Though this noninteraction was the customary way, I did find stimulating exceptions at Duke in the nascent program in Science, Technology and Human Values, under whose auspices I team-taught with historians who shared their culture with me, and I mine with them.

On other occasions when the two cultures encountered each other socially, usually by an accident of intermarriage (love trumps intellectual pursuits) that brought invitations from across the divide, the humanists often put the engineer on the defensive with questions relating to nuclear accidents, planned obsolescence, airplane crashes, and the like. The engineer was expected to explain and defend his (we were almost all male then) humanity. Professional interaction could take place through team-teaching, which was encouraged but not significantly rewarded by the administration, and committee work, which was expected but often avoided. On committees there was parity, and equanimity, and they provided opportunities for the two cultures to interact, usually without overt posturing, but mostly in small groups where biases and stereotypes were dismissed; members were considered the exceptions that proved the rule. Such opportunities enabled engineers and nonengineers (including scientists) to get to know each other and to realize that there were not only differences but also similarities in their approaches to issues and tasks.

It was in part questions from my faculty colleagues that prompted me to write an extended essay about engineering design and failure, seeking not only to comprehend better why some designs end in failure but also to communicate to readers that failures were, if not inevitable, understandable in the context of the big picture of the interrelated engineering and social processes. To make the resulting book, *To Engineer Is Human: The Role of Failure in Successful Design*, more attractive and accessible to nonengineering readers, I used a good number of first-person anecdotes and alluded to and quoted freely from nursery rhymes, poetry, and literature. This approach seems to have attracted a wider readership than an engineering monograph or textbook would have and helped overcome some of the usual barriers between Snow's two cultures. One prejudice of my humanities colleagues that seemed to persist, however, was the belief that engineers could not write, and I was asked more than a few times whether my wife—an English major in college, then a writer of fiction and a teacher of writing—had in fact ghostwritten the book for me. I have certainly learned a great deal about writing from my wife, who has been my first reader and saved me from countless grammatical and rhetorical embarrassments, but she has always had enough of her own thinking and writing to do.[13]

Through college courses in philosophy, literature, and history I had learned to enjoy reading and, by extension, learned the rudiments of writing. It remains a curricular requirement that engineers become familiar with more than technical matters. Unfortunately, the same is not universally true of humanities majors. Generally speaking, there is no requirement that an English major (or a science major, for that matter) take a course in engineering, or a history major one in technology, "Science for Poets" courses notwithstanding. Even with the realization of C. P. Snow's vision about the spread of the Industrial Revolution around the globe, many potential community, national, and world leaders receive no formal education in serious scientific or technical subjects, even though they may be expected to play crucial roles in deciding the future of their district and our planet. These days, more than ever, engineering solutions to problems rooted in changes wrought by the Industrial Revolution must be informed by more than technical calculations and concerns. A calculus of values must accompany any mechanics of function.

It is not necessary to have formal exposure to a subject to appreciate it and master its principles, however. Over the course of time many, but far from all, of my colleagues on the opposite side of Science Drive have revealed that their latent interests in and knowledge of things scientific and technical are remarkably broad and sophisticated. They are as attracted to the beauty of structures as an engineer can be to the structure of poetry. Though it may not have been easy for them to express such extradisciplinary interests in graduate school or among their disciplinary colleagues decades ago—or even to an interloper before they knew him very well—many of them do so enthusiastically among kindred spirits today. C. P. Snow did acknowledge that there were such exceptions to the two-cultures rule among his own colleagues, but he chose to emphasize the norm, who are often some of the most influential and effective members of their respective cultures. In fact, today there are countless shuttle diplomats between the sciences and the humanities, and it is reassuring to know that there are no physical laws or rules of rhetoric to prevent the crossing of the divide.

If the two cultures of a half century ago were the sciences and the humanities, are the two cultures of today the sciences and engineering? Do scientists understand engineering, and vice versa? There is clear evidence that individual scientists do, for they in effect do engineering when they invent and design hypotheses or apparatus to test them. And there are engineers who not only understand science but also engage in it, often under the guise of "engineering science." But the overall cultures of the sciences and engineering can be as disparate as those that Snow observed between the sciences and humanities. While there are scientists who look down on engineering and engineers who dismiss science as of no practical value, in an age of apparent climate change and other global issues, it is incumbent upon both cultures to see the importance of the other in defining and solving the problems of the planet.

Similarly, it behooves scientists and engineers to be connected with the cultures of the humanities and social sciences. Solutions to global problems must take into account matters of humanity and society, which is not always done while in the throes of experimentation and modeling, first to understand the technical issues and then to deal with them. The goal, after all, is not science and engineering for their own sake, but for the sake of the planet and its inhabitants. We all should strive to be of one culture—and not talk past, down to, or over the heads of each other. Ironically, it may be the very fact that potentially devastating planetwide problems like climate change can by definition affect everyone everywhere that will lead to an all-inclusiveness in the world's efforts to solve them. There can be little doubt that these are not times for the global scientific, engineering, economic, political, and public policy communities to separate themselves into competing cultures. They can best unite when they understand each other's disciplines and their essential roles in contributing to the whole.

12

Uncertain Science and Engineering

There can be no absolute certainty about a scientific prediction that an asteroid will strike the Earth, that a hurricane will strike New Orleans, that an earthquake will strike Los Angeles, or that it will rain on tomorrow's cookout in Peoria. Even with the steady improvement in computer models on which such predictions are based, they represent only probable outcomes, albeit ones with increasingly explicit indications of their likelihood. Thus, as a hurricane moves over the Gulf toward Mexico and Texas, we can follow on television its projected course, complete with color-coded graphic expectations that it will strike one location rather than another. And when the hosts of the midday cookout in Peoria check tomorrow's hourly forecast on Weather.com, they might find that the chance of precipitation is given as 40 percent at noon and 10 percent at three o'clock. They can at least have the luxury of playing the odds and leaving the afternoon cookout as scheduled, with the contingency plan that they will do the grilling in the carport and serve the hamburgers and hot dogs inside the house.

By definition, a likely outcome is never truly a sure thing, as anyone who has invested in the stock market, bet on a horserace, or followed the polls leading up to an election knows. Thus, an expectation of a particular future for Earth's climate is a scientist's best educated guess that it will happen, based on theories, models, data, and judgment. In the case of hurricane predictions, the graphics that advance across the television screen toward landfall are often a composite of the results of several different predictions by several different groups of experts. The capsule summary of these predictions is a kind of average educated guess, which changes as conditions change. As most people have noticed, the predictors, like the weather forecasters, can be uncannily accurate in their forecast, but they can also be surprisingly far off target.

According to Natalie Angier, "the words 'science' and 'uncertainty' deserve linkage in a dictionary." This is probably a good idea, but not one likely to be adopted by lexicographers. People who make and read dictionaries prefer definitive answers to their questions about words and their meanings, and the conventional wisdom is that science is sure. In fact, that is often the way its findings are reported. Admitting too explicitly a wide uncertainty in a prediction, a concept, or a definition can make us uncomfortable. Yet the reality is that there can be great uncertainty in what we want to be certain. In recounting her research in preparation for writing her book on the "beautiful basics of science," Angier recalled how "scientists talked about the need to embrace the world as you find it, not as you wish it to be." But no one, scientist or not, seems to want to embrace a hot and dirty planet full of risk and uncertainty.[1]

As much as the prediction of an earthquake is a scientific one based on scientific evidence, records, and theories, what to do in preparation for the Big One devolves upon policy makers and engineers, whose job it is to give us what we wish for. Merely knowing that the scientific fact of Newton's First Law—that a body in motion tends to continue along a straight path and that a body at rest tends to remain at rest until acted on by an outside force—applies does nothing to protect life and property during an earthquake. The problem that the engineer faces is how to keep the structures built on shaky ground from being set in such motion that they are shaken apart and fall on people. Generally speaking, the responsibility of the scientist qua scientist ends with the warning, which is where the responsibility of the engineer begins. (Of course, scientists can always take off their scientist hats and don those of engineers at that point.) Maps and charts of seismic proclivity based on scientific observations of historical data and known faults form the basis for engineering considerations of how much concrete or steel should be used in the buildings designed for a region prone to earthquakes. But if the scientists are not absolutely certain in their predictions and warnings, how certain can the engineers be of the efficacy of their solutions to the problems that they tackle? And where do they start?

Just as scientific predictions come with a degree of uncertainty, so engineering solutions can be fully reliable only to the extent that they take into account predictive uncertainties and uncertain outcomes. Structural engineering, the discipline that most concerns itself with mitigating the effects of earthquakes, must deal constantly with problems full of uncertainties. Indeed, this fact is emphasized in an incisive definition frequently cited by practitioners: "Structural engineering is the art of assembling materials whose properties we do not fully understand into arrangements we cannot fully analyze to support loads we cannot fully predict—and to do so in a convincing enough fashion so that the public has complete confidence in the resultant structures."

But structural or any other kind of engineering can go only so far in exercising its judgment. Earthquakes, hurricanes, and other potentially catastrophic events categorized by Richter, Saffir-Simpson, and other specific scales of intensity come under the scrutiny of politicians as well. When scientific predictions tell us all that there is such and such a probability of a Richter 8 earthquake occurring in California over the next hundred years, it becomes a matter of public policy (one hopes informed by engineering judgment), about whether to design and construct bridges, buildings, and other structures to withstand such a strong ground motion or whether essentially to take a chance with public safety and design for only a Richter 7 quake, playing the odds that the Big One will not strike before another political opportunity arises to reevaluate the situation and refine the prediction—and to check the budget to see what level of further engineering and protection is affordable at the time.

The powerful earthquake that struck San Francisco in 1906 provided a benchmark against which the Golden Gate Bridge was designed and built decades later. The magnitudes of the forces involved in the famous earthquake were inferred from the nature of the damage done to buildings, and the towers and approach spans of the bridge were made to be capable of withstanding similar ground-shaking-induced horizontal inertial forces. Little more was known

Engineers can ensure that the Golden Gate Bridge is not damaged by an earthquake or other terrestrial forces, but the earthquake itself, unlike an asteroid strike, cannot be prevented. The bridge is shown here packed with pedestrians celebrating the fiftieth anniversary of its opening.

about earthquakes and how structures withstood them at the time. Three years after the Golden Gate Bridge opened, the El Centro earthquake struck in Imperial Valley. It was the first earthquake on which data was collected as the event was occurring. Over the years, more sensitive and sophisticated earthquake monitoring devices were developed, along with theories on the causes and characteristics of the notoriously unpredictable natural events. When the San Fernando earthquake struck in 1971, recording instruments were in place to measure its magnitude, which registered 6.6 on the Richter scale.

The San Fernando earthquake damaged almost seventy bridges that had been designed to be "quake-proof," and so the California Department of Transportation (commonly known as Caltrans), reconsidered its design criteria. Understandably, the structure of the Golden Gate Bridge was analyzed anew in light of the knowledge about earthquakes and their effects that had developed since it was built. Engineers found that the towers of the bridge proper were capable of withstanding a Richter 8.25 quake, but some of the approach structures could be vulnerable to a quake measuring no more than Richter 5.5. Structural modifications were made accordingly. In addition, plans were put forward to install seismic recording instruments at critical locations on the bridge in order to measure the magnitudes of forces and movements it would experience during an actual earthquake, whenever one might strike. This kind of information would be extremely helpful in guiding a post-earthquake inspection of the structure to determine what damage might have occurred and to assess the ability of the bridge to continue to carry traffic safely.

For reasons having to do partly with lack of funds, the instrumentation was not yet in place when the 1989 Loma Prieta earthquake struck, damaging the San Francisco–Oakland Bay Bridge and closing it to traffic until repairs could be made. This major disruption in the area's infrastructure emphasized the importance and urgency of installing computerized seismic instrumentation systems on the Golden Gate and other bridges. Not only would this instrumentation provide instantaneous measurements of earthquake ground motion and structural response, but also the data collected would help scientists advance their understanding of the nature of earthquakes themselves and of the scientific prediction of their occurrence. Clearly, in this case, advances in engineering and science proceeded side by side, with a step forward by one enabling the other to take perhaps two steps forward.[2]

A recent study sponsored by the U.S. Geological Survey and conducted by the Working Group on California Earthquake Probabilities estimated there to be a 99 percent probability that an earthquake with a magnitude of at least 6.7 on the Richter scale would occur in that state within thirty years. During that same period, the probability that an earthquake with a magnitude of at least 7.5 would strike Southern California was reported to be 46 percent. Furthermore, the survey also estimated a 10 percent chance that the offshore Cascadia Subduction Zone would experience an earthquake of magnitude 8 to 9 somewhere along its 750-mile length during that same thirty-year period. Not all the news was bad, however, for among other results of the study was that the probability of activity in the Elsinore and San Jacinto faults was less than previously thought. The multidisciplinary working group of scientists and engineers that conducted the study considered information from the fields of seismology, geology, and geodesy. They provided a bewildering number of predictions, with a wide range of probabilities and magnitudes, for designers of budgets and bridges to take under advisement.[3]

Such predictions prompt engineers to design schemes to upgrade the bridges and buildings located in the areas of expected earthquake activity. Typically, separate estimates are prepared detailing how much it would cost to retrofit existing structures so that they can withstand either a Richter 7 or a Richter 8 quake, but which to choose for the retrofitting? As was the case with the Golden Gate Bridge, a common compromise is to go ahead with a lower level of protection now, with the promise of revisiting the issue at a later time. But to bring structural protection up from Richter 7 to Richter 8 standards can be enormously expensive, disproportionately more than the difference between the simple numbers 7 and 8. Since engineers typically do not control infrastructure budgets, which course of action to follow is not their decision alone. Engineers might advocate one strategy over another, but they as well as any other interested parties will know that to go with the more expensive option will mean that other infrastructure projects will not be able to be carried out in the city, region, state, or nation. It might, for example, come down to a choice between making bridges able to resist large earthquakes and making levees able to resist large floods. Budgetary pressures will favor the less expensive solution, and it is these pressures that often dominate political decision making.

Such was the case in New Orleans, where a Category 5 hurricane had been expected to strike someday, but levees were constructed to withstand only Category 3 conditions. Even in the wake of Hurricane Katrina, it was clear that the mere possibility of an even stronger hurricane hitting the city was not a sufficiently compelling argument for spending the additional money to rebuild the levees and bring related storm protection equipment up to the higher standard. There are enormous political pressures against spending the money to engineer defenses to withstand the worst possible scenario, in large part because of the low probability of such a scenario occurring.

Budgetary considerations aside, how certain are engineering solutions to protect us against natural hazards? As Katrina demonstrated, the existence of a deliberately designed "engineered barrier" system of levees, pumps, canals, and the like is no guarantee that a city such as New Orleans is adequately protected. Exceptionally high tides and strong winds combined to produce unanticipated high storm surges that pushed water over levee walls, resulting in erosion of the back slope. This in turn allowed the concrete walls to tip back and allow even more water to breach the levee, a chain of events apparently unforeseen as being probable. Even if foreseen, it was not considered the fatal flaw that it proved to be. Engineered barrier systems like levees are only as good as each individual section, something that should be obvious. However, as with anything designed, understanding how any single part of it could fail is a key consideration in ensuring its success. The facts that the New Orleans levees were inadequately designed, poorly maintained, and overly relied upon combined to lower the probability that they would function adequately when tested by the likes of Katrina. One would hope that the lessons learned from that storm, as well as those from Hurricane Gustav three years later, will ensure more robustly redesigned hurricane protection systems. Nevertheless, a review panel convened by the Army Corps of Engineers shortly after Katrina occurred found that even the improvements made since that time left the city and its residents with an "unacceptable" risk should it be struck by another storm of similar ferocity and circumstances.[4]

As Katrina also demonstrated, the protection of the residents of a city like New Orleans cannot depend solely on an engineered barrier system that is susceptible to undetected or unsuspected flaws of design, construction, and maintenance. The possibility of the levees being topped or giving way must always be considered, and plans to evacuate the population have to be designed as surely as do the earthen and reinforced-concrete levees that are supposed to keep the water out. One of the tragedies of New Orleans was that an effective emergency preparedness plan was apparently not in place when Katrina struck. Even if it were, not implementing it would be tantamount to not having it.

Science and engineering, working in tandem, can do only so much to predict and prevent the terrible consequences of low-probability events, whether in the natural or in the made world. When the U.S. airline industry had so few accidents over the course of years, it faced the paradoxical situation of not having enough accidents to analyze to improve safety. In other words, because safety was at such a high level, it was difficult to raise it to an even higher level. We learn best from past mistakes and accidents how to reduce and avoid future mistakes and accidents, and thereby increase safety. In the absence of accidents, the next best thing for the airline industry to do is to collect data on "accident precursors," those minor and individually ignorable events that could lead to catastrophe. But the obvious way to collect information on precursors is to rely on the "voluntary disclosure" of incidents that would generally be known only to the airlines and personnel experiencing them. Since the Federal Aviation Administration has developed a reputation for fining airlines that reported errors relating to, say, missed inspections, it is unlikely that such accident precursors will be reported if they do not have to be. An independent panel convened to look into airline safety issues recommended, among other things, that the FAA adopt a no-penalty voluntary disclosure practice similar to what was in place and working in other government agencies, like the Nuclear Regulatory Commission and the Environmental Protection Agency. The secretary of transportation ordered the FAA to implement the panel's recommendations.[5]

In the case of natural disasters, there should be little impediment to reporting precursors associated with hurricanes, earthquakes, and volcanic eruptions. Scientists know that tropical storms can grow into hurricanes, and so tropical storms are tracked. There are centuries of experience and scores of scientific hypotheses about what signals a potential earthquake or eruption, and scientists are eager to have any precursor data that can help them test their hypotheses. In the case of a large asteroid potentially striking Earth, the precursor information relates to the orbit of the near-Earth body; tracking it over the course of years or decades can be the most reliable means of providing as specific a warning as possible of an impending cataclysmic event. And if and when a man-made defense is set up to avert or mitigate a natural disaster, any lessons learned from an encounter that does not destroy the entire planet will prove invaluable for identifying weaknesses in the defense and for the improved design of future defenses.

As one Corps of Engineers colonel said about the upgraded New Orleans system of protective levees, "You can't out-engineer Mother Nature." Since it is extremely expensive to devise and implement protective measures against major natural onslaughts, it becomes a matter of national public policy to decide which potential catastrophes to concentrate resources upon. Do we spend more money protecting the Gulf Coast against hurricanes or the West Coast against earthquakes? Do we give the highest priority to a long-shot asteroid strike, designing and stockpiling the necessary equipment to set in motion the machinery to deflect a gigantic incoming rock as soon as it is detected and determined to have a high probability of doing extensive damage? The list of questions could go on and on.[6]

For historically infrequent events, such as nuclear war and earth-shaking volcanic eruptions, there is naturally a paucity of hard data on how often they can be expected to occur in the future and how many lives they may claim. Yet extremely rare but potentially devastating events could threaten the very existence of the human species and so need to be studied. According to John Garrick, who has spent his career thinking about how to analyze and quantify such catastrophic events:

> These are the risks that often don't get taken seriously and yet, they are the kind of risks that could greatly compromise or even terminate life. Many of these risks are beyond known human experience and generally tend not to raise much action from nations and their leaders. How can we get to the truth about risks that are rare and catastrophic and may be even irreversible unless anticipatory actions are taken? How can we possibly manage these risks if we don't really know what they are?[7]

Garrick believes that "rare catastrophic events will most likely be the major threat to a life-sustaining planet for the centuries to follow and that proper analyses will result in better management of such events." But how do we decide which risks are to be given priority and how they and others are to be "managed"? Do we worry more about tsunamis or volcanic eruptions? Or should we focus more on terrorist attacks of the kind that occurred in New York and at the Pentagon on September 11, 2001? We can get a great amount of help in dealing with such questions by employing the methods of risk assessment, especially in the form known as quantitative risk assessment or probabilistic risk assessment, which involves assigning numerical probabilities to imaginable events, scenarios, and effects.

According to Garrick, in order to estimate the amount of risk and damage associated with an event, we must be able to answer three questions:

> 1. What can go wrong?
> 2. How likely is that to happen?
> 3. What are the consequences if it does happen?

It is by systematically answering these questions, however tentatively, that risk can be quantified, even if only probabilistically, and thereby provide some basis for comparing the apples and oranges of hurricanes and earthquakes, tsunamis and global climate change.[8]

Knowing the probability of something happening—and our vulnerability to its occurrence—is not enough, however, if we wish to mitigate the consequences of what might occur. This is where what is known as risk management comes into play. One way of managing risk is to lower the probability of an unwanted event, like the breach of a levee or the collapse of a bridge, from even happening. This can be done by more careful and robust engineering of designed systems themselves, but it cannot be done so easily for potential natural onslaughts of unprecedented (and unanticipated) magnitude, extent, and kind. For man-made threats, like terrorist attacks involving explosives, we might reduce the vulnerability of our infrastructure by producing more hardened defenses, but how hard would be considered hard enough? Should we plan for a nuclear attack? And what about biological weapons of mass destruction?

Historical case studies of the occurrence and response to actual accidents and disasters of a lesser degree can provide invaluable experience on which to build. Where there is little direct experience, as is the case with extreme risks like asteroid strikes, we might draw upon key generic lessons and extrapolate from them. But who will do the extrapolating? The American Society of Civil Engineers, in its vision of what the profession would be called upon to do in 2025, expected its practitioners to be "managers of risk and uncertainty caused by natural events, accidents, and other threats."[9]

Regardless of who manages risk, devising engineering strategies to mitigate the effects of a catastrophic event itself comes with a degree of uncertainty. Engineering typically involves the creation of something new, but when this means conceiving, designing, and implementing a scheme on perhaps an extra-global scale, there is little precedent and much risk. This is not to say that such ambitious attempts are doomed to failure, for the ultimate success of the Apollo program in landing astronauts on the Moon and bringing them back to Earth is a striking counterexample to the false dictum that we *must* fail before we succeed. There were failures and setbacks along the way to the first Moon landing, but the first time the whole system was tested through to touching down on the Moon and blasting off again and returning to Earth, it worked. In fact, subsequent missions to the Moon, which included the "successful failure" of the aborted

Apollo 13 mission, were all carried out without loss of life. Indeed, aside from the real-time drama associated with Apollo 13, Moon excursions quickly became so predictably successful that the media deemed them not to be newsworthy.

The space shuttle program also involved a wide range of new technology and a great deal of risk. As risky as each launch-to-landing excursion was, after the first few they became as unremarkable as men on the Moon had become. Managers involved in the shuttle program had estimated the probability of "failure with loss of vehicle and of human life" to be one thousand times less likely than the estimate of the engineers, who put it at about one in one hundred. The first two dozen shuttle missions were successful, and the failure during launch of Challenger put the historic probability of failure at one in twenty-five. After improvements were made in both equipment and procedure, shuttle missions resumed. In spite of the dangers presented by foam shedding from the external fuel tank, shuttle launches continued until some of that foam damaged the leading edge of the wing of the shuttle Columbia, which disintegrated upon reentry. After this second failure, the historic probability rate stood at about one in fifty-six. With the successful return of Atlantis from its mission to upgrade the Hubble Space Telescope in May 2009, the empirical rate became one in sixty-two. Estimates of risk understandably evolve with experience, optimistic managers notwithstanding.[10]

Launching rockets carrying massive payloads of explosive materials to be landed, mounted, and operated on asteroids in order to redirect their orbit away from Earth clearly involves new and risky technology. However, there is no reason to believe that such a mission could not be successfully executed, despite the high risk factor. According to Garrick, writing about not only a large-asteroid impact but also about catastrophic events generally and our responses to them, "there is seldom enough data about future events to be absolutely certain about when and where they will occur and what the consequences might be. But 'certainty' is seldom necessary to greatly improve the chances of making good decisions."[11]

Most important, Garrick believes that to "make good decisions on the management and control of catastrophic risks," they have first to be quantified. He recognizes the difficulty of quantifying "something so rare as a modern-day global famine; an astronomical event leading to complete or partial extinction of life on Earth; a hundred- or thousand-year severe storm, earthquake, or volcanic eruption; a terrorist attack that can kill tens or hundreds of thousands of people; or a climate change that could lead to total extinction of life on Earth." However, the alternative to not trying to think about and deal rationally with such rare and terrible events is to leave our most powerful weapon in its scabbard. Inaction can result in "complacency on the part of society about being able to control such risks." Without an appreciation for the real risks involved, however unlikely the threatening events may be, we cannot make informed decisions about how to engineer and execute our defense.[12]

One cosmic catastrophe that scientists have discussed for some time is the possibility if not the probability that, as our sun matures as a star over the course of billions of years, it will eventually grow into a red giant before collapsing into a white dwarf. In this transition period, the sun may engulf and "eventually desiccate Earth, leaving it hot, brown and uninhabitable." There is nothing definite about this prediction, since it depends upon so many assumptions about the gravitational and orbital interactions among the bodies of our solar system. However, a "bold piece of astronomical engineering" has been proposed whereby the Earth could escape this fate. The scheme involves "nudging Earth with a large asteroid arranged to pass nearby periodically." Over the course of a billion years or so, this could expand Earth's orbit to a safe distance from the sun, thereby putting it out of harm's way.[13]

No matter how little experience there might be with a rare event—one that might happen only once in a few centuries or even once in several millennia, or maybe even only once in the life of a star—it may be possible to establish a probability, which can be taken as somewhat synonymous with credibility. In the case of an asteroid hitting the Earth, even prehistoric examples known through geological evidence can be used to infer a probability. This crude estimate can be enhanced by means of probability theory to incorporate other more recent impact evidence, such as craters in remote locations. For example, asteroids with an impact energy equivalent to between

10,000 and 100,000 megatons of TNT, which could produce a blast area on the order of ten thousand square miles—an area about the size of Vermont—can be expected to strike the contiguous United States about once in every 81,000 years, thus presenting a small but real risk. Obviously, where such an asteroid strike might take place would have a great effect on how many lives would be at stake. The probability of a strike in population areas of varying density can also be calculated.[14]

By employing scientific principles and numerical data to quantify risk, priorities can be set and problems solved. It will not be a question of whether scientists or engineers can save the planet from catastrophic events but rather one of cooperative contributions of the best of all sectors of knowledge and technique working together toward a common and commonly beneficial end. According to the summary of National Academies reports on climate change, for example, "the task of mitigating and adapting to the impacts of climate change will require worldwide collaborative input from a wide range of experts, including physical scientists, engineers, social scientists, medical scientists, business leaders, economists, and decision-makers at all levels of government." We are all in it together, and we all must contribute to solving our common problems. Some of our contributions will be as observers and predictors, some as lawmakers and policy makers, some as interpreters and problem solvers. It is in the latter categories that engineers can be expected to make their most important contributions, as they have so many times in the past.[15]

13

Great Achievements and Grand Challenges

In the mid-1990s, when the end of the millennium was in sight, there was a considerable amount of looking back at what had happened and what had been accomplished, especially during the twentieth century. The National Academy of Engineering convened a committee to identify the greatest engineering achievements of the age, and the resulting list provided a record of how much improvement there had been in the quality of life in the developed world. The "twenty engineering achievements that changed our lives" included, in the order of importance that the committee voted them: electrification, the automobile, the airplane, water supply and distribution, electronics, radio and television, the mechanization of agriculture, computers, the telephone system, air-conditioning and refrigeration, highways, spacecraft, the Internet, imaging, household appliances, health technologies, petroleum and petrochemical technologies, lasers and fiber optics, nuclear technologies, and high-performance materials.[1]

The list provides clear evidence of the essentially interdisciplinary nature of so much engineering achievement. The electrification of America, for example, has clearly been the success that it is because of a collection of what might be called subachievements. Each of these may have been significant in its own field, but none would be as meaningful without the others. The whole is clearly greater than the sum of the parts. Electrical engineers provided the theoretical foundations whereby power transmission over long distances was made economical; mechanical engineers contributed their expertise to making efficient and reliable generators; chemical engineers to developing lubricants and coolants for the machines; civil engineers to building dams for hydroelectric plants; petroleum engineers and mining engineers to ensuring supplies of oil and coal for fossil-fuel plants; nuclear engineers to providing an alternative source of energy to generate power for the grid; materials engineers to providing effective conducting and insulating materials; structural engineers to designing reliable steel transmission towers. No great achievement is wholly the province of a single engineering discipline, but each can take pride in its contributions to the greater good.

Of course, the scientific discovery of electricity itself was a sine qua non for its distribution. However, knowing that something exists and obeys certain laws does not in itself deliver that thing to rural farmhouses. While some scientists evidently believed that the National Academy's celebration of the engineering achievement of electrification neglected to acknowledge the theoretical and experimental physicists who had laid the scientific foundations for it, they could not alter the fact that, as we have seen, things sometimes can be accomplished by engineers without a full understanding of the enabling theories or restricting laws. But, at the same time, the chairman of the selection committee, H. Guy Stever, reminded us that "engineering has advanced physics by developing instruments and equipment for research." In many ways, as we have also seen, engineering and science support and advance each other.[2]

The word *achievement* suggests completeness, a goal reached, a task finished. The great engineering achievements of the twentieth century do indeed represent in a metaphorical sense the culmination of great adventures for engineers. But engineering feats and exploits are really never-ending. Even the greatest achievements comprise indistinct milestones passed on the road to the future. (Exactly when and where did electrification actually take place?) Still, the list of achievements has been hailed as a virtual blueprint for bringing forth the advancement of conditions in less developed countries. It is the nature of engineers to play a central role in such development, for it is the nature of engineering to effect change.

Engineering, as engineers especially know all too well, is a continuing process. It is a journey with frequent stops, much backtracking, and many redirections, but never a truly final destination. What is engineered may, momentarily, be admired for what it is. But the individual components of engineering achievements are like leaves that have fallen into a stream, where they are carried by swift waters from the fresh springs of the past to the still lakes of the future. There, they precipitate into the silt of technology, perhaps someday to be unearthed by an industrial archaeologist.

As much as electrification was a great achievement, the way it was accomplished in the United States has left us with a distribution grid that is complicated, fragile, and vulnerable. In the wake of severe hurricanes or winter storms, hundreds of thousands if not millions of citizens can be without electrical power for days, weeks, or months. The overhead power lines on which we rely are exposed to falling tree limbs and accumulated ice, which can lead to their being severed. In anticipation of a major storm capable of doing major damage, power-line workers from throughout a region mobilize, converge on the target area, and await the inevitable. Broken power poles are replaced, fallen lines restrung, blown transformers exchanged for new ones, and customers made happy by having their power restored. How much more sensible in a country of great achievements would it be to bury all power lines and thereby minimize, if not virtually eliminate, power outages associated with severe storms? As an aesthetic by-product, there would no longer be unsightly lines sagging between leaning poles or the need for tree-trimming services that butcher boughs to allow cables to pass.

But even burying all the nation's electric power lines out of harm's way would not completely eliminate interruptions of service. On November 9, 1965, the lights suddenly went out for as many as thirty million people in the northeastern United States and parts of Canada. People were stranded in subways and elevators, traffic lights did not work, and there was considerable confusion and inconvenience. The cause of the problem was traced to a relay switch, which was supposed to act like a circuit breaker, preventing an overload of the system. However, when the breaker was installed, it was set to be activated by a lower voltage than necessary, so when a slight surge tripped it, it set off a concatenation of power plant shutdowns. A seemingly trivial component of the international electrical grid, incorrectly installed, was its undoing. On August 14,

2003, not only the northeastern states but also some in the Midwest were hit with another great blackout, which occurred suddenly. About fifty million people in the United States and Canada were left without power (and air-conditioning) on a very warm summer day. Because it was so hot, high-voltage power lines had stretched and drooped more than usual, and one in Ohio is believed to have touched a tree that had been allowed to grow too close to the line. This caused a short circuit, which in turn caused hundreds of power plants feeding the grid to shut down, leaving the vast region without electricity.[3]

Western states have not been immune to interruptions in service. In 2000 and 2001, California and other states were subject to "rolling blackouts," which were caused not by malfunctioning equipment but by a combination of regulatory failures, market manipulation, accelerated demand, and a diminished supply of hydroelectric power due in part to drought conditions in the Northwest. When California partially deregulated the power supply industry, prices rose out of control. Utilities, which bought power from producers and sold it to consumers, found themselves in the position of paying more for the commodity than they could charge for it. This led to bankruptcies and public bailouts of failed utilities. The experience demonstrated how inextricably intertwined technology, government, and markets can be. Just as science is a never-ending quest to uncover the mysteries of the universe, so engineering is the never-ending pursuit of a better system, including how the nuts and bolts will interface with the dollars and cents and the supply and demand.[4]

No achievement, however great, is without its limitations. The automobile was second on the list of great engineering achievements of the twentieth century, but what is an automobile? Unlike the early-twentieth-century driver of a Stanley Steamer, who had to be prepared to water his machine when it was thirsty and fix it when it was sick on a rutted rural road, today's car owner need hardly look under the hood of what has become an easy-to-operate and most reliable and healthy, if not diet-conscious, thing. When did the automobile as achievement become a fait accompli? Was it when the toolbox disappeared from the running board? Or was it when the electric starter replaced the hand crank? Perhaps it was when the system of roads—also collectively recognized in their own right as one of the great achievements of the past century—became paved and numbered. Or was it when there was no longer a need to carry cans of emergency gasoline? Would the automobile be what it is today without the infrastructure of roads and filling stations and repair shops? The achievements captured under the rubrics *automobile* and *highways* really connote those of entire interrelated systems.[5]

Like all engineered systems, that of the automobile and its infrastructure is ever evolving. Who but the collector or first-car buyer in a developing nation today wants a car without a radio, intermittent windshield wipers, cruise control, or cup holders? Need the machine called an automobile have included all of these things before it qualified as a great achievement? Even if we could specify a checklist of what it takes for an automobile to achieve majority, we might be dissatisfied with our present model as soon as the next model year comes out with the latest driving aids and comfort gadgets.

It might take someone who read *Motor Trend* and attended the annual auto shows of the 1950s to remember the differences between 1955 and 1956 models. As a rule, from year to year the automobile changed slowly, but the cumulative effect of a hundred years of small, gradual changes—with a sizable one now and then—has made a big difference. And thus it is with a lot of engineering achievements that we sweepingly summarize in a word or two. Was the steam engine an achievement of the seventeenth or eighteenth century? Will the World Wide Web ultimately be associated with the twentieth century or with the twenty-first, when it can be expected to be truly accessible to people around the globe, in developed and developing nations alike?

Although engineers always seek to make everything better, they cannot make anything perfect. This basic characteristic flaw of the products of the profession's practitioners is what drives change and makes achievement a process rather than simply a goal. Understanding this essential fact enables us to speculate with some degree of confidence about future improvements of even the greatest of engineering achievements. By identifying what is still wanting in today's technology, we can predict with considerable confidence what will be standard in tomorrow's. This is not to say that we can see exactly what form the future will take, for we recall that engineering differs from mathematics in that it makes no claim to unique solutions. Whether videocassette recorders would one day use the VHS or Betamax format was not at all obvious at the outset of those technologies in the mid-1970s, even to many of those closest to the development of the systems.[6]

Consider a car from 1950. If I remember correctly, it typically had manual transmission, no power steering, a flat split windshield, and one-speed windshield wipers. Yet it was perfection to a teenager. New-car buyers were generally extremely pleased with the latest styling and technology, but they were also very much aware of their car's limitations. Learning to drive a stick-shift—coordinating clutch, accelerator, gearshift, steering wheel, feet, and hands—was not a trivial task. Getting moving up a steep hill after stopping for a red light was tricky. The automatic transmission has made driving a car much less of a test of mechanical aptitude and foot-arm coordination. Power steering removed the need for weight training to be able to turn into a tight parking space. Wraparound windshields eliminated the obstructive posts in the field of view. Intermittent windshield wipers eliminated the need to turn a knob on and off while driving through a drizzle. These innovations grew out of the recognition by inventors, manufacturers, and consumers alike that there was room for improvement.

To predict how the automobile will continue to change and improve over the next century we need only look at what annoys us about it today or what features we wish it had or think it should have. Design is effectively proactive failure analysis, so if we perform a conceptual survey of the early-twenty-first-century automobile, we can predict with some degree of confidence what changes are likely to occur over the coming years and decades. But it is easier to pinpoint technological faults than to predict precisely how and when they might be satisfactorily corrected without introducing new problems. Some drivers might wish that their cars were quieter, and so the hybrid or all-electric vehicle might seem to be a godsend. However, blind pedestrians have come to rely upon the sound of the internal combustion engine to warn them of an approaching car. The introduction of silent models has led to the idea of incorporating synthetic engine noises into electrics, so that the visually impaired could hear them coming. But what of the hearing impaired person who is not looking at approaching traffic?[7]

In an interview, the science-fiction writer Ray Bradbury once was asked, "If you could eliminate one invention from the last 100 years, what would it be?" He answered, "The automobile," because it had "killed two million people." Bradbury, who said that he had never driven a car, likened the highway carnage to "a major war" and lamented the fact that "we're not paying any attention to it." Designing safer cars (and safer highways) should clearly be a goal of engineers in the twenty-first century. What these cars (and highways) will look like and exactly what protective features they will have should be of less interest in this case than the accomplishment of the goal. Style and features will follow from the process of working to achieve that.[8]

A less deadly problem with today's automobiles is that they lead drivers into unknown territory. We get lost. Global positioning satellite technology, which can obviate this, is increasingly being incorporated into new generations of cars. But not knowing our coordinates is only one way of getting lost. I heard recently of a woman who was listening to a book-on-tape while driving around Philadelphia. She got so engrossed in the narrative that before she knew it she was in Ohio, three hours from home. (I can believe this story, because once, while listening to a John Grisham novel in pre-GPS days, I missed the I-85 exit off I-95 and drove for half an hour toward Rocky Mount, North Carolina, instead of toward my home in Durham, before realizing my mistake.) The development of smart vehicles, which can operate like airplanes on automatic pilot, could mean that we would be able to read books and watch television and maybe even sleep while cruising down the highway without ending up in Ohio, Rocky Mount, or elsewhere.

The airplane and its associated infrastructure of airports and ticketing schemes have come a long way since the historic flights of the Wright brothers, and the ensemble that goes under the rubric *airplane* certainly warrants recognition as a great achievement. Airplanes may be crowded with passengers and the skies with planes, but there is still room for improvement in aviation, especially in the area of safety of small aircraft and their operation. As good as the safety record of the commercial airline industry is, there is the occasional accident that in retrospect often was clearly avoidable. Early in the new millennium, there were two aircraft accidents that were particularly shocking: an Air France Concorde crashed shortly after takeoff from Paris, and a Singapore Airlines Boeing 747 broke up after hurtling down a closed runway in Taipei, Taiwan. In each case, the accident resulted from the plane striking something on the ground. Such incidents were to lead to procedures for ensuring that open runways are clear of debris and that those under construction are clearly marked "closed" when that is in fact their status. Though the Air France and Singapore Airlines disasters may not change much of the overall physical appearance of airplanes and runways, accidents of this kind influence the way airports are maintained and used, which is an implicit part of the achievement of air travel. In the case of the Concorde, the accident played a large role in leading to the retirement of the entire fleet.

Without faults or accidents, actual or imagined, there might be little driving change in large technological systems. The Concorde was nearing the end of its originally intended design life when the Air France accident happened. Yet, given the outstanding performance, safety record, and physical condition of the aircraft before that incident, there seemed to be little reason to take it out of service. Had the Paris accident not occurred, it is likely that aging Concordes would have continued to fly, perhaps until a serious accident of another kind occurred. Though the Paris crash had nothing to do with the fact that the Concorde could fly at supersonic speeds, the incident would affect the way supersonic aircraft were perceived for some time.

The conventional wisdom might have had it that to be listed among the greatest engineering achievements, a technology should have been "perfected." But, as we have seen, there is no such thing as perfection in artifacts, for engineering is the art of compromise and of continuing betterment. The sleek Concorde had a relatively low passenger capacity because the plane's fuselage was small in diameter, a structural necessity because the cabin had to be highly pressurized in order to carry people at almost sixty thousand feet—about twice the altitude at which conventional jets typically fly—which in turn was necessary to reduce drag and conserve fuel. All other things being equal, flying very fast requires more fuel than flying more slowly, which obviously makes the journey more expensive. In a technological system, each part of the whole necessarily affects every other part.

The achievement of the airplane as we know it would be nothing without the infrastructure of airports and the affiliated systems dealing with aircraft maintenance, airline reservations, air traffic control, and baggage handling. All aspects of airport operations have long relied heavily on computers, but the physical-labor-intensive chore of baggage handling remains perhaps the last frontier. It was going to be crossed and conquered once and for all in a big way as part of building the largest airport in the world. Denver residents and visitors had long appreciated the convenient close-in location of the city's Stapleton Airport, which dated from 1929, but by the mid-1970s it was clear that a larger facility was needed. Stapleton's near-downtown location then turned from advantage to liability, since the area's growth and established land use around the airport left little room for expansion. Furthermore, Stapleton's parallel runways were too close together to allow their simultaneous use in poor weather, thus causing delays in landings and takeoffs that affected airline schedules well beyond Denver. Given the situation, the city decided to design and build a new airport that would overcome the limitations of the old and be state-of-the-art.[9]

Making any change in an established infrastructure can have profound implications. The new Denver International Airport was to be located on an enormous tract of open land about twenty-five miles from downtown—more than three times farther out than Stapleton—which meant considerably longer driving distances and times for most Denver residents and visitors. To ameliorate this, a new highway system was planned for construction along with the airport. But even new roads can be plagued with maddeningly long and slow commutes. And riding in airplanes taxiing around an airport can be equally frustrating. I have been on many a flight that landed ten or fifteen minutes ahead of schedule, only to deplane ten or fifteen minutes late because of the distance the plane had to cover and the active runways it had to cross to get to the gate—which was still sometimes occupied by another plane.

Denver International covers fifty-three square miles, making it larger than the (very large) Chicago O'Hare and Dallas–Fort Worth airports combined. Indeed, DIA is located on "the largest piece of real estate dedicated to commercial aviation on earth." Still, it was designed to minimize taxiing time. Its main runways are laid out in a pinwheel pattern, with two main ones aligned roughly north–south and two east–west. Instead of locating the pairs of parallel runways beside each other, they are on opposite sides of the terminal, staggered so that one end of each runway is relatively close to the terminal. Since planes land and take off into the wind, the optimistic expectation was that there would always be a preferred runway nearby, within easy taxiing distance, so that no matter what the wind conditions, planes could take off going away from the terminal. Incoming aircraft would use the corresponding runway that would enable them to land toward the terminal, thus also minimizing their taxiing distance and transit time to the gate. However, according to one critic, this expectation was a "myth" that did not hold up under certain weather conditions and busy periods, requiring planes to taxi as much as three miles. (Laying out the terminal and parking structures to reduce driving and walking time for travelers also proved not to be fully successful. The main terminal—a vast interior space surmounted by a structurally daring fabric roof that is supposed to evoke mountain peaks but instead, one observer has noted, "rises out of the high plains like an extraterrestrial circus big top"—is a visually striking design, but the elevated roads and parking garages surrounding it unfortunately block the view of the Front Range of the Rocky Mountains to the west.)[10]

One of the most discussed features of the new airport was what was to happen not above- but belowground. Denver International's baggage handling system, promised to be "the largest such system in the world," was to be wholly automated, with bar-coded pieces of luggage tracked by laser scanners and deposited into and carried everywhere by more than 3,500 individual hopper cars (450 of which were large enough to accommodate skis and golf bags) driven by the magnetic forces imposed by linear induction motors mounted along

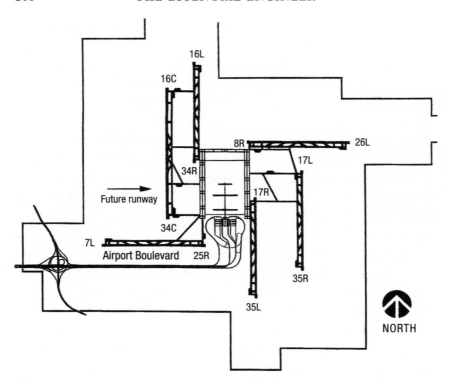

A plan for runways for the Denver International Airport project had them laid out in a pinwheel arrangement. This was to minimize taxiing time for both arriving and departing aircraft.

the tracks. The system's thousands of electric motors and photocells were connected to hundreds of computers and ancillary equipment by fourteen million feet of wire. Such a large and complex system understandably requires extensive testing and fine-tuning before being ready for prime time. However well it may have performed in private tests, it failed miserably and embarrassingly when the local media were first invited to view it in action. Baggage was catapulted errantly from hoppers, and the area was littered with clothing from popped-open luggage. The system never did work properly, delaying the opening of the airport because there was no backup plan. When Denver's new airport eventually did open, it was with a much more modest and more conventional baggage handling system. The designers discovered that it is easier to reach for something than to achieve it.[11]

All of the engineering achievements that were identified as being among the greatest of the past century leave room for improvement. Air-conditioning and refrigeration are among the more domestic of the achievements of twentieth-century engineering, but at exactly what stage they became so is also hard to say. Refrigerators especially came a long way in the second half of the century. At about mid-century, refrigerators were streamlined like automobiles on the outside, but inside most contained a freezer compartment no larger than a breadbox. Its accumulation of caked-up ice had to be defrosted regularly and with no little mess left on the floor. The frost-free freezer was as welcome an addition to the kitchen as the self-cleaning oven has been. Making ice cubes became trouble-free as the process was automated behind the freezer door. But even opening the door to retrieve some ice cubes was seen by some inventor-engineers as something to be improved upon, and so the ice cube dispenser was introduced into the front of the door. How can refrigerators be developed further? Clearly, they can be improved in the way items are stored inside, for it is certainly inconvenient to have to grope behind the turkey carcass to find the cranberry sauce for a late-night Thanksgiving snack. Does this mean that future home refrigerators will acquire some of the features of vending machines, whereby the push of a button—or perhaps a voice command—brings the desired item close at hand?

Air-conditioning works wonderfully when there are wonderful air conditions. But when too many people crowd into a room on a hot and humid day, despite its being air-conditioned the room too often feels too crowded, too hot, and too humid. In such circumstances, the greatness of the engineering achievement might understandably be called into question. And how is it that the cooling, heating, and ventilating systems of new and old buildings alike are so difficult to control? The United Nations building in New York is notorious for its problematic climate control. Achieving temperature uniformity in the 1952 structure "ranks up there with world peace as a noble, if unlikely, goal. Some rooms, notably the General Assembly and the basement, are frigid. Others feel distinctly tropical." Part of the problem appears to be that the building's interior has been modified so much since it was constructed that thermostats have become separated from the rooms whose climate they are supposed to control.[12]

Like the little girl with a curl right in the middle of her forehead, when "modern" climate control is good it can be very, very good, but when it is bad it is horrid. Along with agricultural mechanization, air-conditioning seems to be among the most unglamorous of the achievements to make the celebrated list, and in addition seems to be the most finicky. Or is it that one central air-conditioning system is expected to be all things to all people at a meeting? Who has not felt too cold when sitting directly across the table from someone who felt too hot? Who has not had to decide between the draft and the relief? For all of the technological progress made in the field since air was fanned over a block of ice, well-controlled air-conditioning seems to remain one of the great open problems of mechanical engineering. While it is a problem that pales in importance next to dealing with global warming, it should not surprise us to see meeting rooms with individualized comfort control among a list of great achievements at the end of the twenty-first century.[13]

All achievements, engineering and otherwise, are relative to their time and circumstances. When we look back over the last century, we see clearly that technological progress was made, and we rejoice in it. But this is not to say that engineers have said the final word on automobiles, airplanes, air-conditioning, or any of the other great achievements in mechanical or any other branch of engineering. Rather, these accomplishments have set down ever-evolving standards against which achievements of the next century will be measured. Engineers of the last millennium, as engineers of all millennia, did the best they could with what they had to work with at the time. Engineers of the new millennium can be expected to do no less. The tools they use may be different, but the goal they strive for is the same: improvement.

It has been almost fifty years since I first learned to use a slide rule, one of the great innovations of the seventeenth century. The slide rule worked well for what it did under the circumstances, but it clearly limited an engineer's reach. During my career, the electronic calculator buried Napier's bones and the digital computer has led to computer-aided design, computer-aided manufacturing, and a host of related computer-based technologies. But the more we use these new tools, the more we discover their limitations and their faults. What made the twentieth century different from the nineteenth, which itself saw the rapid rise of the railroad, the telegraph, the steamship, and a host of other great innovations, was the rapidity with which improvements were achieved and diffused throughout society. (Is the personal computer really only a few decades old?) What is most likely to characterize the twenty-first century is an even faster rate of change from the good to the better. The engineering of the best is always yet to come; when it has apparently arrived, it will already be fleeting.

Engineering achievements do not come without environmental cost. For every one of the twenty innovations celebrated by the National Academy of Engineering, a contrarian or critic could find an environmental fault. Electrification brought fossil-fuel-burning power plants and high-voltage transmission lines, which some see as a blight on the landscape and others as a health hazard. The automobile emits greenhouse gases. Airplanes do also, and they produce considerable noise pollution, too. Water supplies are laced with fluoride, which some consumers believe is insidious. Electronics manufacturing produces heavy-metal by-products. Radio and television can pollute young minds, or so it is said. Agricultural mechanization has claimed many a young life and limb. Computer use leads to carpal tunnel syndrome. Cell phones may cause brain damage. Air-conditioning reduces an automobile's fuel efficiency. Et cetera, etc., &c. There is always room for more R&D.

These are important lessons to remember when engineers look to tackling and are looked to for tackling the global problems that threaten planet Earth. Proposed solutions will have shortcomings and outright flaws that one hopes will be caught before there is full implementation of a novel scheme. And there will be bumps in the road to achievement. Engineers know this, and it is why they like to advance slowly and methodically, working with their scientific partners in research and development first on the laboratory scale before attempting progressively larger-scale demonstrations on the way to full-scale implementation. What works on the laboratory bench top does not always work out in the field, however, and it is important to understand this before investing time and money that might be better spent by going in another direction. In tackling global problems this is especially important, because an errant effort to help the environment could result in an environmental disaster of global proportions.

Early in the twenty-first century, there was a heightened awareness of the downside of technology, especially the way it can adversely affect the environment on a grand scale. But as much as the inadvertent harmful by-products of technological achievement might be blamed for everything from local smog to global warming, it is also solid engineering and enlightened public policy that will be necessary to reverse the negative effects and bring forth new achievements for a new time. There have been notable reversals of environmental damage, such as the cleanup of the air over cities like Pittsburgh and London, whose industry and coal burning at one time made their visibility range seem more reduced than that of Beijing before the Olympics. Indeed, in order to reduce the impact of its chronic pollution on athletes and visitors, China restricted traffic and ordered certain factories closed in the month before the 2008 summer games. Such decisive measures can be effective, but the global scope of problems faced today calls for new approaches and new and more lasting solutions.[14]

The National Academy of Engineering anticipated this in convening another committee—comprising inventors, engineers, and scientists—which was charged with identifying "grand challenges" for twenty-first-century engineering. According to a press release, "rather than focusing on predictions or gee-whiz gadgets, the goal was to identify what needs to be done to help people and the planet thrive." The resulting challenges fell into "four themes that are essential for humanity to flourish—sustainability, health, reducing vulnerability, and joy of living." These themes helped group the unranked list of fourteen challenges:

- make solar energy affordable
- provide energy from fusion
- develop carbon sequestration methods

- manage the nitrogen cycle
- provide access to clean water
- restore and improve urban infrastructure
- advance health informatics
- engineer better medicines
- reverse-engineer the brain
- prevent nuclear terror
- secure cyberspace
- enhance virtual reality
- advance personalized learning
- engineer the tools for scientific discovery[15]

Every such list reflects the nature and diversity of its composing committee. Another committee might have emphasized the development of energy conservation techniques and affordable electric cars; called for the prevention and mitigation of biological and chemical terrorism; and asked for the enhancement of real virtue in the world financial system. No matter what the challenges, listed or not, they will be accepted by engineers armed with tools both classical and modern. No matter what the problem, the engineer's mind and mind's eye will conceive a tentative solution and the computer's memory and brain will help fill out the details to test its viability. This is the way it works for challenges modest and grand.

That is not to say that solutions will come easily. Shortly after assuming the office of secretary of energy, Steven Chu expressed his belief that solving some of our global energy and environmental problems will take breakthroughs worthy of Nobel Prizes. In this category he singled out development work in solar power, electric batteries, and new crops for biofuels. Solar technology, for example, will have to improve fivefold before the challenge associated with it could be considered met. He also expressed the opinion that a revolution in "science and technology," which he uses as a singular noun, would be necessary to reduce the world's dependence on fossil fuels and control greenhouse gas emissions. In other words, the challenges facing scientists and engineers are enormous, and they can be expected to take some wrong turns and hit some speed bumps and potholes on the way to effective solutions.[16]

What has come to be known as financial engineering may be said to have contributed at least in part to the debacle of 2008 involving the banking and investment industries, relating especially to the issuance of and default on so-called subprime mortgages. Among the things financial engineers do is design innovative securities that are intended to spread risk and also design sophisticated mathematical models to track markets and implement strategy. Trading in mortgage-backed securities was based largely on computer models rather than human judgment. But such models can be overly simplistic, incorporating as they do what is graspable by the quantitative finance analysts—financial engineers known as "quants"—and not taking into account the "messy, intractable challenges" of the real world.[17]

After the financial meltdown in the fall of 2008, *Scientific American* identified the quants as "lapsed physicists and mathematical virtuosos" who devised computer models predicting risk. These "rocket scientists and geeks" bore some of the blame, according to the magazine, but so did the Securities and Exchange Commission, which at the request of investment bankers a few years earlier had relaxed debt limits and capital reserves. The models developed in such a climate incorporated unrealistic assumptions about how the markets and their players behaved, which was where "reality and rocket science diverge." One critical economist was reported to have said that the mathematical models of risk used by Wall Street contained "a lot of wishful thinking about house prices." In fact, the true market value of an asset such as a house or stock that you invest in "reflects not only your beliefs about the future, but you're also betting on other people's beliefs," which were not adequately modeled. Thus, according to another economist, the innovative "technology got ahead of our ability to use it in responsible ways."[18]

Not all engineered systems, especially those complicated by challenging ulterior social and economic motives, necessarily evolve for the better. The best of engineering looks at the lifetime of a structure, whether it be a tall building or a financial market. Much of the problem on Wall Street that became evident in 2008 seems to have stemmed from too close attention paid to quarterly earnings reports and other short-term goals. However, like bricks-and-mortar structures, manufacturing businesses, service corporations, and financial institutions need also to address their long-term health through investments in routine maintenance, inspections, and reality checks, as well as in flashy innovation. No engineered structure is designed to be built and then neglected or ignored. Had the list of grand challenges been compiled in the fall of 2008, might it have included a bullet point calling for engineers to help restructure the world financial system?

What will come of the challenges that were laid down will, of course, depend at least in part on how committed scientists, engineers, and governments are to pursuing them intellectually, technically, and financially. It will also depend on what unforeseen developments there might be on the way to the twenty-second century. A list of challenges compiled at the beginning of the twentieth century would not likely have included goals relating to computers, spacecraft, the Internet, lasers and fiber optics, and nuclear technologies. Lists of engineering challenges tend to relate to known problems that have eluded satisfactory engineering solutions. Thus, the list of grand challenges addresses such matters as solar and fusion energy, infrastructure, and nuclear terror, technologies and issues that have been around for decades, if not longer.

The challenges relating to sustainability might be said to focus on distinct aspects of the problem, without consideration of "what level of energy use would be sustainable" on a global scale. That question has been asked by scientists affiliated with the Swiss Federal Institute of Technology, who decided that the answer was a power consumption equivalent to burning continuously twenty 100-watt bulbs, or 2,000 watts total, per individual. People in underdeveloped countries get along with considerably less: the average citizen of Bangladesh gets by with only 300 watts. India is a 1,000-watt society, and China a 1,500-watt one. In the West, Switzerland is rated at 5,000 watts, other European countries at 6,000 watts, and the United States and Canada at 12,000 watts. The scientists who instigated such comparisons organized the 2,000 Watt Society project to pursue the implications of this goal and promote ways of reaching it.

According to a project white paper, in order to achieve a 2,000-watt society "three things are needed: societal decisions . . . technical innovation, and the resolve of every individual to act in an energy-conscious way." A former director of the society, who was trained as a theoretical physicist, was reported to have said that "as a scientist he could see no technical barriers to creating a two-thousand-watt world." He believed that "nothing has to be invented—for an engineer it's not even a challenge." However, he also maintained that it was a "new kind of problem, which cannot be solved with the same recipe as the flight to the moon, or the Manhattan Project. It's a qualitative difference—a paradigm change in the role of science for our society." Unquestionably, society itself will have to play an important role in providing the resources, incentives, motivation, commitment, and will to achieve global goals that are unprecedented.[19]

With societal involvement, meeting some of the grand challenges laid down by the National Academy of Engineering and the 2,000 Watt Society will still require doing engineering if not on a grand scale then in a grand manner. In the final analysis, it will be engineering that possesses the same qualities involved in accomplishing the great achievements of the last century that will be the key ingredient in a solution. True, the new global challenges are qualitatively different from the milestone space and weapons programs, but the problems that engineering faced in solving those problems were themselves thought to be qualitatively different from what had come before. One thing is certain, however: whatever solutions are designed and implemented, they are not likely to be perfect—and this bears repeating—either the first time they are tried, or the second, or the third . . . or the nth time. That is not to say that they will outright fail, but they will need to be assessed, reassessed, and improved upon in an evolving environment of science, technology, and public policy.

Even if we cannot expect perfect solutions the first time, engineering is always up to the challenge of a supposedly qualitatively new problem. And there will always be room for contributions by amateur and independent inventors, as well as by large research and development laboratories—and by individuals who just want to be good citizens. There will be cooperation among engineers, scientists, and medical doctors, but in meeting the grand challenges the participants will in effect all be doing engineering. The up-front verbs used in the bulleted list of challenges are not the scientific verbs of discovery and understanding; they are active calls for creative achievement. They say "engineer" and they challenge engineering.

14

Prizing Engineering

During the 2008 presidential campaign, with gasoline approaching five dollars a gallon in California, candidate John McCain proposed in a speech there that "we inspire the ingenuity and resolve of the American people by offering a $300 million prize for the development of a battery package that has the size, capacity, cost and power to leapfrog the commercially available plug-in hybrids or electric cars." While the amount of the prize might have seemed huge, he emphasized that it represented only "one dollar, one dollar, for every man, woman and child in the U.S.—a small price to pay for helping to break the back of our oil dependency."[1]

Such language can be fine for a political speech, but engineers who might want to compete for the prize would want to know exactly what was meant by the words *development, battery package, size, capacity, cost, power, leapfrog, commercially available*, and others. McCain seemed to anticipate this somewhat, for he elaborated that the winner of the prize "should deliver power at 30 percent of current costs." Presumably, if McCain had been elected president and carried through with his promise, the conditions under which the prize could be claimed would have been spelled out in sufficient detail to satisfy the most precise engineer.[2]

A series of "electrical storage devices" was given the name "battery"—after the word used for an assemblage of artillery pieces, such as the guns on a warship—by Benjamin Franklin. Batteries have come a long way since Franklin experimented with kites and lightning in the eighteenth century, but the technology has evolved very slowly since the mid-nineteenth. Even the high-tech revolution of the later twentieth century did not improve batteries as fast as it did electronic devices. In contrast to the digital storage capacity of a personal computer, which increased over 35,000 times during the two decades following its introduction, the power generated by the batteries that run portable computers has risen only about 10 percent per year. The output of a battery can be increased by increasing its size, but this necessarily makes for a heavier and bulkier power supply. Furthermore, whereas conventional batteries weigh essentially the same whether fully charged or fully discharged, fuel cells grow lighter as they consume their fuel to produce electricity, and so would appear to be preferred.[3]

Senator McCain's promise followed by only days a newspaper story describing a Pentagon-sponsored competition to encourage the "development of new technologies, such as fuel cells, that can provide more juice and weigh less than bulky batteries." Known as the Wearable Power Prize, the contest was motivated by the fact that modern soldiers carry into the field as much as twenty pounds of batteries to run their night-vision goggles, radios, and computers. To win the contest, a device had to weigh less than nine pounds and provide at least ninety-six hours of continuous power. First prize was $1 million, with second and third prizes being $500,000 and $250,000, respectively. These amounts are not likely to cover the cost of research and development for such a demanding new product, but presumably the market and other advantages gained by the winner could make the effort profitable in the long run.[4]

Instead of using the conventional method of soliciting bids for new technology, the prize was instituted in order to reach R&D teams that were outside the "traditional Department of Defense suppliers." The strategy worked, and over one hundred teams, from the corporate, academic, and private sectors, had qualified for the competition that involved a technology that was centuries old. The Wearable Power Prize was won by a prototype portable fuel-cell device developed jointly by DuPont and the German firm Smart Fuel Cell, which specializes in alternative electricity generators for recreational vehicles and boats.[5]

A similar scheme for encouraging new technologies has been successfully employed by the Defense Advanced Research Projects Agency, which had been created in 1957 in the wake of the Soviet

Union's successful launch of Sputnik. DARPA's mission then was to "prevent technological surprise." In recent years, its mission has been expanded to include the creation of surprises as well. The support of challenge prizes is in keeping with such a mission, and one of its most visible is intended to "hasten the development of robotic delivery trucks." Since 2004, the agency has been holding a series of supporting competitions with cash prizes for coming up with the "sensors and software needed to steer a vehicle without a driver at the wheel." As a by-product of the DARPA Grand Challenge, rivalries among research labs "have led to big leaps in terrain-mapping and collision-avoidance technologies." A Super Battery Prize valued at $1 billion—or $10 billion for a still more powerful device capable of storing megawatt-hours—has been proposed but not immediately funded.[6]

The use of prizes to encourage invention and innovation in a specific area is not new. In 1714 the British government established a £20,000 prize for a means of accurately determining longitude at sea. The prize, which would have been worth about $5 million in 2008 dollars, was administered and judged by the Board of Longitude, which proved difficult to satisfy, perhaps because it "received more than a few weird and wonderful suggestions. Like squaring the circle or inventing a perpetual motion machine, the phrase 'finding the longitude' became a sort of catchphrase for the pursuits of fools and lunatics. Many people believed that the problem simply could not be solved." The saga lasted for decades, and the clockmaker John Harrison spent the better part of his professional life pursuing the prize by making a chronometer that would retain its great accuracy regardless of conditions at sea. Since the clock told a captain the exact time in London when the sun was at high noon over his ship, the longitude—essentially the equivalent of the distance from London—could easily be calculated. However, the board proved to be as parsimonious with the prize purse as it was with its approval: to receive just part of the money, Harrison had to jump through hoops to satisfy the board's demands. Finally, after he petitioned Parliament, Harrison's achievement was acknowledged and the balance of the prize awarded to him.[7]

Another famous prize was established in 1919 by the French-born New York hotel owner Raymond Orteig, who offered $25,000 "for the first nonstop aircraft flight between New York and Paris." Many a pilot lost his life in pursuit of the Orteig Prize and the glory that the achievement would bring. The prize was finally won in 1927 by Charles Lindbergh and his thirty-three-and-a-half-hour solo flight in the single-engine airplane, the *Spirit of St. Louis*. In a remarkable interpretation of the engineering concepts of redundancy, reliability, and risk, he had chosen to use a plane with only one engine because he is said to have "believed that the more engines an aircraft had, the greater the possibility of engine failure"! Lindbergh evidently disregarded the benefit side of risk-benefit analysis.[8]

When most scientists and engineers—and laypersons—hear the word *prize*, it is often Nobel Prizes that come first to mind. In his last will and testament, the chemical engineer Alfred Nobel established the prizes to recognize and reward handsomely discoveries, inventions, and improvements in the fields of physics, chemistry, and physiology or medicine—as well as literature and peace. He did not throw down any specific challenge or define any contest, and his lack of specificity left the prizes to be defined and awarded by those who had the most to do with establishing the Nobel Foundation. Two engineers had been named by Nobel to serve as executors of his estate, but like many an engineer they did not take a particularly active interest in nontechnical matters and failed to foresee the public policy implications of setting up a mechanism for administering, judging, and awarding the prizes. The engineer-executors engaged a lawyer, who in turn sought counsel from scientists of his social acquaintance. The rest is, of course, history.[9]

Since "discoveries, inventions, and improvements" could be interpreted as including engineering achievements involving the application of physics and chemistry, what have come to be known as the Nobel science prizes could have been established as honoring science *and* engineering, at least, rather than becoming the virtually exclusive province of science. When early Nobel Prizes were awarded to scientists for scientific accomplishments of long standing, there was some debate about whether Nobel's intentions were being carried out. But the scientific establishment prevailed, and engineers essentially lost their chance. Scientists can be inventors, of course, and engineers discoverers, but by the middle of the twentieth century it was clear that the Nobel Prizes in physics and chemistry were considered truly science prizes, being rarely awarded to engineers or for engineering achievements.[10]

Professional engineering organizations, including the National Academy of Engineering, approached the Nobel Foundation with a proposal to endow an engineering Nobel Prize akin to the now familiar economics prize, which was not mentioned in Nobel's will. Officially designated the Bank of Sweden Prize in Economic Sciences in Memory of Alfred Nobel, and offered for the first time in 1969, the economics prize is thus not a Nobel Prize in the strictest sense, but that distinction is lost in the annual fall announcements of the "Nobel Prizes." However, it was not only the engineering profession that sought to associate another prize with the Nobels; for mathematicians and others were also interested in joining the exclusive club. Understandably, the Nobel Foundation saw adding more associated prizes as potentially diluting the status and cachet of the original (plus economics) prizes.

With no hope of a formal association with the Nobel Prizes, in 1988 the National Academy of Engineering established its own prize, complete with a comparable honorarium. The $500,000 Charles Stark Draper Prize is awarded annually to recognize "individuals whose outstanding engineering achievements have contributed to the well-being and freedom of all humanity." (The first Draper Prize was awarded in 1989 to the engineers Jack S. Kilby and Robert N. Noyce, who three decades earlier had independently invented and developed the integrated circuit, which formed the foundation of the consumer electronics industry. Kilby, one of the few engineers to win the Nobel Prize in physics, received it in 2000 for the invention of the integrated circuit. Had Noyce not died in 1990, he likely would have been co-recipient of the prize.) As much as the academy has tried to get the press and public to think of the Draper as an "engineering Nobel prize," the annual announcement of its winner has been largely ignored by the popular media. The Nobel Prizes, like journalism's own Pulitzer Prizes, are the ones that get front-page treatment. Virtually all other prize announcements, whether associated with engineering or some other excluded profession, are used as fillers, if they are mentioned in the newspaper at all.[11]

The introduction of multimillion-dollar challenge-specific prizes, which by the very nature of the challenge and their cash value qualify them as significant news, has made it even less likely that prizes like the Draper—and the engineering achievements that they honor—will get the recognition that they deserve. Other large-money retrospective prizes, such as the $500,000 Lemelson-MIT Prize, which "honors outstanding mid-career inventors dedicated to improving our world through technological invention and innovation," have had an equally difficult time competing with the Nobels for attention from and recognition by the general news media.[12]

Not only have specific contests and competitions become noteworthy, but also they have become institutionalized, being described by *The Economist* as "one of the most intriguing trends in philanthropy: promoting change by offering prizes." The X PRIZE Foundation was established for just such a purpose. The original X competition was the $10 million Ansari X PRIZE, named after the foundation board member and Texas telecommunications entrepreneur who funded it through an insurance policy. The competition to launch the first privately financed spacecraft to carry "three adults to an altitude of 100 kilometers, twice within two weeks," was won in 2004 by SpaceShipOne, which was designed by a team led by the aerospace engineer Burt Rutan. His second-generation design, the "reusable spaceliner" SpaceShipTwo, is a key component in the space tourism firm Virgin Galactic's plans to carry civilians into space at a cost of $200,000 per ride. The spaceliner will be taken aloft on a mothership designated WhiteKnightTwo and will be launched into its suborbital trajectory from the air. For those who want to experience orbiting Earth, the firm Space Adventures will transport passengers willing to pay the megaprize price of $30 million for the tourism package that puts them on the International Space Station for a period of time. As of 2008, half a dozen paying customers had been up there and done that.[13]

The Google Lunar X PRIZE, announced in 2007, challenges contestants "to land a rover on the moon that will be able to travel at least 500 meters and send high resolution video, still images and other data back home." Those achievements would earn the winner $20 million, with an additional $5 million being a kind of bonus for roving ten times as far or for transmitting images of a lunar lander or other artifact from the Apollo program. The prize was to be held out for about five years, with the possibility of its being extended with a lower purse, presumably because technology will have advanced by that time and so have altered the playing field.[14]

The Automotive X PRIZE was to go to the designers of a safe car that gets one hundred miles per gallon while meeting "tough emission standards" and being affordable. The last achievement requires the submission of a business plan demonstrating that ten thousand of the vehicles could be manufactured at a "reasonable cost." When the competition was announced with draft guidelines, it was widely criticized. Some environmentally conscious observers asked: Why just one hundred miles per gallon? Why not two hundred? Others faulted the challenge for not attempting to separate automobile travel from dependency on oil. They felt that it would not contribute to alleviating global warming. The organizers of the prize defended their guidelines as the result of a year of listening to "scientists and automotive experts." (One can only hope that there were some experienced engineers among the experts.) Following this process, goals were set that were believed to "balance what is achievable, what will make a difference to humanity, and what will create a playing field where small teams and big auto companies can compete on the same level."[15]

About a year after the draft guidelines were released, the Progressive Insurance Company joined with the X PRIZE Foundation to define the final form and name of the Progressive Automotive X PRIZE. The one-hundred-mile-per-gallon objective remained in place for the "international competition designed to inspire a new generation of viable, super fuel-efficient vehicles." The $10 million prize purse promoting "revolution through competition" was unveiled at the 2008 New York International Auto Show. The competition involves both mainstream and alternative vehicles: the former were to have at least four wheels and carry at least four passengers two hundred miles; the latter were to carry at least two passengers one hundred miles. The mainstream winner was to receive three times as much of the prize money as the alternative winner. At the time of the announcement in the spring, sixty teams had indicated their intention to compete for the prize; by the end of the summer, twice that many had expressed interest. Cross-country races scheduled for 2009 and 2010 were to help determine the winners.[16]

A vehicle competition that might satisfy critics of the Automotive X PRIZE is named Escape from Berkeley, which has been described as "maybe the world's most eco-friendly motor race." The 2008 petroleum-free road rally from Berkeley, California, to Las Vegas, Nevada, was to be a six-hundred-mile affair, but a snowstorm at the almost ten-thousand-foot-high Tioga Pass in the Sierra Nevada added a two-hundred-mile detour to the course. The rally drew creative alternative-fuel-burning vehicles of many kinds, including motorized bicycles, to compete for the relatively modest $5,000 first prize. One entrant was a pickup truck whose engine runs on the combustion gases given off by scrap wood burned in the truck's bed. Other fuels included vegetable oil, kudzu, corncobs, and phone books. Entrants began with the equivalent of only one gallon of their fuel of choice aboard and had to "find or scavenge" beside the road any further fuel needed to complete the course. The race is sponsored by Shipyard Labs, which has been described as an "open-air garage" where many of the entrants work on their vehicles out of shipping containers. According to the founder of Shipyard, the purpose of the race is to promote "creative thinking about alternative energy." He wishes to change approaches to energy independence "from an engineering problem to art." What may not be acknowledged, however, is that solutions to engineering problems always involve an element of creativity akin to art, and if artists take up the challenge, they will effectively be doing creative engineering.[17]

Not all contests are for such visible, news-making, and well-defined achievements as fuel-efficient cars or independent space travel. The Buckminster Fuller Challenge, which carries a $100,000 prize, is intended "to support the development and implementation of a solution that has significant potential to solve humanity's most pressing problems in the shortest possible time while enhancing the Earth's ecological integrity." Entries of "design science solutions" to entrant-defined problems are judged on how well they exemplify Fuller's "trimtab principle," whereby "small amounts of energy and resources precisely applied at the right time and place can produce maximum advantageous change." One entry in the 2007 competition involved "an eco-motel with an enclosed greenhouse to grow vegetables, fish and livestock, and an eco–housing development to provide for 1,000 residents with food and energy self-sufficiency." Its designer was inspired by Fuller's own 1971 "Old Man River's City" proposal for East St. Louis, Illinois. This consisted of "a massive housing project that would have held 125,000 occupants, in the form of a circular multi-terraced dome."[18]

The National Academy of Engineering administers the Grainger Challenge Prize for Sustainability. The first challenge was to develop "innovative solutions for removing arsenic from drinking water that is slowly poisoning tens of millions of people in developing countries." It was motivated by the situation in Bangladesh and neighboring parts of India, where over the years an "estimated 10 million tube wells were built with international aid to provide an alternative to bacteria-tainted surface water. Unfortunately, these wells frequently tap into aquifers contaminated by arsenic." To win the prize, a filtering system had to be "affordable, reliable, easy to maintain, socially acceptable, and environmentally friendly." The winning systems, which were announced in 2007, required no electricity to operate and met or exceeded local government regulations regarding arsenic. The first prize, the $1 million Grainger Challenge Gold Award, went to a system that had been under development as many as eight years before the prize was even announced. It employs principally common and locally available materials, including buckets, river sand, wood charcoal, and brick chips.[19]

Not all challenges have an obvious humanitarian goal like providing safe drinking water. In 2000 the Canadian mining company Goldcorp sponsored a contest with prizes totaling $500,000 for the identification of sites where drilling would improve the yield of one of the firm's gold mines. As a result of the company's making available to contestants the geological data for its Ontario mine, previously unidentified drill sites were discovered, which increased the annual yield from the mine by a factor of ten, to 500,000 ounces of gold. Some years later, "in an unusual approach to research and development," another Canadian mining firm, Barrick Gold, announced a $10 million prize competition open "to any scientist, researcher or inventor who can increase the amount of silver the company recovers from a mine in Argentina."[20]

Whereas Barrick was accustomed to recovering 80 percent of the gold from its mines, it was achieving only a 6.7 percent yield of silver from the Argentine source. Rather than relying on its small R&D staff to come up with a solution, the company cast a wide net to see what kinds of innovative approaches might come from the "global scientific community," which presumably included engineers. Details about the mine and information relating to the silver problem were posted on the Web, which made it available to experts and amateurs alike. Anticipating that not all ideas submitted would be considered good practice for the environmentally sensitive operation of silver mining, the company's review committee was expected to "filter out the crazy stuff pretty quickly."[21]

A competition of modest purse but potentially earthshaking implications has been sponsored by the Planetary Society. Its focus is asteroid 99942 Apophis, whose provisional name was 2004 MN4. This rocky object, believed to be "a primordial relic from the formation of our Solar System," orbits the sun in such a way that it comes close to Earth every decade or so. At one time Apophis was rated at level 4 on the Torino Impact Hazard Scale, and it was thought to have as much as a 2.7 percent chance of striking our planet. It will pass close—but not dangerously close—to Earth in 2017. In 2029, it is expected to come so close that it will be nearer than a geostationary satellite and will be able to be seen with the naked eye. Its next encounter with Earth will be in 2036, when it might strike our planet. One estimate has put the chance of this happening at 1 in 45,000, which are frighteningly small odds for such a potentially cataclysmic event. In order to get a better estimate of how close to Earth the asteroid actually will come—and whether it will be on a collision course—it is desirable to take accurate measurements of the roughly thousand-foot-diameter space rock over an extended period of time, but well in advance of crunch time. This was the basis for the challenge set down by the Planetary Society, which hoped to "catalyse the world's space agencies to move ahead with designs and missions to protect Earth from potentially dangerous asteroids and comets."[22]

The challenge was to design a "mission which would launch, rendezvous and collect enough data in time for governments to decide in 2017 whether or not to mount a mission to deflect the asteroid off its current course." The prize was valued at £25,000, which is a lot less than the eighteenth-century longitude prize would be worth today, but the challenge attracted entries from around the world. The winner was SpaceWorks Engineering, which is based in Atlanta; its winning scheme involves a spacecraft called Foresight that will rendezvous and follow the asteroid Apophis for about ten months, collecting and transmitting essential data. An accurate determination of the body's center of mass, for example, will reduce uncertainties in its predicted orbit.[23]

The hope is to launch Foresight in 2012, thus allowing time to make preparations, if deemed necessary, to "build a spacecraft powerful enough to move the space rock out of harm's way." If such a mission were called for, it should ideally be launched before 2025, so that the cumulative effect of a small nudge of the asteroid's orbit would be sufficiently great for it to miss Earth a decade thereafter. Since such careful planning ahead cannot be made to prevent the poorly understood phenomena of earthquakes and hurricanes, whose locations and paths cannot be nearly as accurately predicted, an asteroid strike might be "the one truly preventable natural disaster." And, of course, if it is to be prevented, it will take some spectacular engineering. But scientists and engineers, working together with governments and public policy makers, know that they are up to the challenge of out-of-this-world engineering.[24]

In some more down-to-earth cases of saving our planet, the engineering may have already been done, in the form of developing wind turbines, heat pumps, and other clean and efficient ways of producing and redistributing energy. The need would be for a commitment to use them in a comprehensive and determined way. In 1997, the Danish Ministry of Environment and Energy sponsored a contest to promote innovation in renewable energy use. The challenge to each community was to put forth a plan demonstrating "how it could wean itself off fossil fuels." An interested engineer identified the island of Samsø, population 4,300, as a self-contained community that might be considered in thermodynamic terms almost a closed system of sorts. With the cooperation of the island's mayor, the engineer prepared a scheme and entered Samsø in the competition. The plan was selected as the winner, and Samsø became Denmark's "renewable energy island." Unfortunately, no prize money accompanied that designation, nor did Samsø receive any special tax incentives or government grants.[25]

Samsingers, as residents of the island are known, were implored to participate in the experiment "as a new kind of social relation" and an opportunity "to work together on something they could all be proud of." People began "thinking about energy," and coming up with ways to reduce fossil fuel use "became a kind of sport." Oil and gas furnaces were replaced with straw-burning ones and heat pumps, solar energy was used to warm water, and wind turbines were constructed. By the spring of 2008, barely ten years after the island won the contest, it was operating over twenty large wind turbines, eleven on land and ten offshore. The cost of the machines was as remarkable as their size, with the land-based turbines costing $850,000 and the sea-based ones about $3 million each. Some of the machines represent the investment of a single individual; others were the collective investment of shareholders. All received a return on their investment based on the sale of wind-generated electricity to utilities, and all expected to recoup their capital outlay in about eight years.[26]

Perhaps the plethora of challenge prizes that have been proposed and offered in recent years will prompt individuals, governments, foundations, and international bodies to issue further challenges, including those that will induce some of the best minds in invention, engineering, science, and citizenship to tackle truly global problems. Regardless of the existence of prizes, there will no doubt be plenty of scientists and engineers interested in working on such problems, whose solutions themselves can be very personally rewarding. Except for those who are independently wealthy, however, being a problem solver takes outside money and other resources, not only to compensate for one's time but also to provide the materials and equipment to observe, calculate, experiment, design, and build whatever devices and systems might be found necessary to save the planet. With an increasing understanding of each other's distinguishing capabilities, scientists and engineers are likely to come together and work as the team that they naturally should be.

Notes

CHAPTER I. UBIQUITOUS RISK

1. Jim Yardley and Andrew Martin, "More Candy from China, Tainted, Is in U.S.," *New York Times*, Oct. 2, 2008, p. A14; "Tainted Chocolates Found in Hong Kong," *New York Times*, Oct. 6, 2008, p. A11.

2. David Barboza, "China's Tainted-Food Inquiry Widens Amid Worries over Animal Feed," *New York Times*, Nov. 1, 2008, p. A8; David Barboza, "Melamine Is Discovered in More Eggs from China," *New York Times*, Oct. 30, 2008, p. A17; Edward Wong, "More Lawsuits Are Filed over Tainted Milk in China," *New York Times*, Oct. 31, 2008, p. A13; Yardley and Martin, "More Candy from China"; Gardiner Harris and Andrew Martin, "U.S. Blocks Products with Milk from China," *New York Times*, Nov. 14, 2008, p. A18; Andrew Jacobs, "China Pledges New Measures to Safeguard Dairy Industry," *New York Times*, Nov. 21, 2008, p. A10; David Barboza, "China Plans to Execute 2 in Scandal over Milk," *New York Times*, Jan. 23, 2009, p. A5.

3. Barboza, "Melamine Is Discovered."

4. U.S. Food and Drug Administration, "FDA Finalizes Report on 2006 Spinach Outbreak," news release, March 23, 2007; Bina Venkataraman, "6 More States Report Illnesses from Tomatoes," *New York Times*, June 13, 2008, p. A21; Bina Venkataraman, "As Outbreak Affects 1,000, Experts See Flaws in Law," *New York Times*, July 10, 2008, p. A13; Annys Shin, "All Tomatoes Cleared of Salmonella Risk, Officials Say," *Washington Post*, July 18, 2008, p. D1; Bina Venkataraman, "Salmonella Strain in Jalapeños Is a Match," *New York Times*, July 22, 2008, p. A22; see also Bina Venkataraman, "Amid Salmonella Case, Food Industry Seems Set to Back Greater Regulation," *New York Times*, July 31, 2008, p. A17; Elizabeth Weise, "Salmonella Outbreak Linked to Jalapenos Appears Over," *USA Today*, Aug. 28, 2008, http://www.usatoday.com/news/washington/2008-08-28-cdc-salmonella_N.htm.

5. Gardiner Harris and Pam Belluck, "New Look at Food Safety After Peanut Tainting," *New York Times*, Jan. 30, 2009, p. A17; "Hazardous Peanut Butters," *New York Times*, Jan. 30, 2009, p. A28; Gardiner Harris, "Peanut Product Recall Took Company Approval," *New York Times*, Feb. 3, 2009, p. A13; Michael Moss, "Peanut Case Shows Holes in Food Safety Net," *New York Times*, Feb. 9, 2009, pp. A1, A12.

6. "Safer Salad," editorial, *New York Times*, Aug. 28, 2008, p. A26.

7. Elisabeth Rosenthal, "New Trend in Biofuels Carries New Risks," *New York Times*, May 21, 2008, p. A6; "Europeans Reconsider Biofuel Goal," *New York Times*, July 8, 2008, pp. C1, C5.

8. Kenneth Chang, "In Study, Researchers Find Nanotubes May Pose Health Risks Similar to Asbestos," *New York Times*, May 21, 2008, p. A18.

9. "Action Urged over Nanomaterials," *BBC News*, Nov. 12, 2008, http://news.bbc.co.uk/2/hi/science/nature/7722620.stm; "Fear of Nanohazards," *ASEE Prism*, Sept. 2008, p. 21.

10. Kevin Sack, "Guidelines Set for Preventing Hospital Infections," *New York Times*, Oct. 9, 2008, p. A21; Shannon Brownlee, *Overtreated: Why Too Much Medicine Is Making Us Sicker and Poorer* (New York: Bloomsbury, 2007), p. 52; Institute of Medicine, *To Err Is Human: Building a Safer Health System* (Washington, D.C.: National Academies Press, 2000).

11. See, e.g., http://www.planecrashinfo.com/cause.htm; Jeff Bailey, "American Cancels Flights to Allow a Reinspection," *New York Times*, April 9, 2008, p. C7.

12. "Emil Konopinski, 78, Atomic Bomb Scientist," obituary, *New York Times*, May 28, 1990, p. 42; Harold M. Schmeck Jr., "Apollo 11 Crew, Unlike Predecessors, Will Be Quarantined on Return from Space," *New York Times*, May 27, 1969, p. 29.

13. Dennis Overbye, "Swiss Particle Accelerator Deemed Safe," *New York Times*, June 21, 2008, p. A9; Dennis Overbye, "Earth Will Survive After All, Physicists Say," *New York Times*, June 21, 2008, http://www.nytimes.com/2008/06/21/science/21cernw.html; Douglas Birch, "Scientists: Nothing to Fear from Atom-Smasher," Associated Press, June 30, 2008; Roger Highfield and Harry Wallop, "Scientists Receive Death Threats over 'End-of-World' Experiment," *Telegraph .co.uk*, Sept. 8, 2008, http://www.telegraph.co.uk/scienceandtechnology/science/sciencenews/3351252/Scientists-receive-death-threat-over-end-of-world-experiment.html; see also "The Large Hadron Collider: Countdown," *Scientific American*, http://www.sciam.com/report.cfm?id=lhc-countdown; Dennis Overbye, "Fingers Crossed, Physicists Are Ready for Collider to Roll," *New York Times*, Sept. 8, 2008,

p. F2; William Booth, "Smashing Idea," *Washington Post*, Sept. 11, 2008, p. C1.

14. Associated Press, "Damage to Atom Smasher Forces 2-Month Halt," *Los Angeles Times*, Sept. 20, 2008, http://www.latimes.com/news/science/la-sci-collider21-2008sep21,0,3810552; Associated Press, "An Explanation of Problems with the Large Hadron Collider, World's Largest Atom Smasher," *Chicago Tribune*, Sept. 20, 2008, http://www.chicagotribune.com/news/nationworld/sns-ap-eu-particle-collider-q,0,3499175.story; Paul Rincon, "What Happened to the Big Bang Machine?" *BBC News*, Sept. 20, 2008, http://news.bbc.co.uk/2/hi/science/nature/7627631.stm; Alexander G. Higgins, "One Bad Electrical Connection Sparked Collider Shutdown," *USA Today*, Oct. 6, 2008, http:// www.usatoday.com / tech / science / 2008-10-6-lhc-shutdown-cause_N.htm; Mark Henderson, "Large Hadron Collider to Stay Switched Off for a Year," *Times Online*, Feb. 10, 2009, http://www.timesonline.co.uk/tel/news/uk/science/article5701040.ece.

15. David Morrison, "Impacts and Evolution: Protecting Earth from Asteroids," American Philosophical Society, Autumn General Meeting, Philadelphia, Nov. 13–15, 2008; Kenneth Chang, "Huge Meteor Strike Explains Mars's Shape, Reports Say," *New York Times*, June 26, 2008, p. A16.

16. William J. Broad, "When Worlds Collide: A Threat to the Earth Is a Joke No Longer," *New York Times*, Aug. 1, 1994, p. A1; Anahad S. O'Connor, "Astronomers Adopt Doomsday Index," *New York Times*, July 27, 1999, p. F4.

17. Broad, "When Worlds Collide," pp. A1, A12.

18. Arthur C. Clarke, "Killer Comets Are Out There. Now What?" *New York Times*, Aug. 14, 1994, p. A15; B. John Garrick, *Quantifying and Controlling Catastrophic Risks* (San Diego: Academic Press, 2008), pp. 79–80, Chapter 4, and Appendix C. Note that a kilometer is about six-tenths of a mile.

19. Http://en.wikipedia.org/wiki/Tunguska_event; David Hawksett, "Tunguska Incident," *Astronomy Education*, http://www.astronomy-education.com/index.php?page=135; "Incident at Tunguska," *Time*, May 6, 1985, http://www.time.com/time/printout/0,8816,967554,00.html; Paul Rincon, "Fire in the Sky: Tunguska at 100," *BBC News*, June 30, 2008, http://news.bbc.co.uk/2/hi/science/nature/7470283.stm.

20. John Johnson, "U.S. Not Prepared for Possible Asteroid Strike, Group Says," *Los Angeles Times*, July 5, 2008; see also "Sandia Supercomputers Offer New Explanation of Tunguska Disaster," news release, Sandia National Laboratories, Dec. 17, 2007.

21. Steve Chesley, Paul Chodas, and Don Yeomans, "Asteroid 2008 TC3

Strikes Earth: Predictions and Observations Agree," NASA Near-Earth Object Program News, Nov. 4, 2008, http://neo.jpl.nasa.gov/news/2008tc3.html; Ashley Yeager, "Great Balls of Fire," *Nature News*, Oct. 8, 2008, http://www.nature.com/news/2008/081008/full/news.2008.1158.html.

22. Yaeger, "Great Balls of Fire"; http://neo.jpl.nasa.gov/stats.

23. Clarke, "Killer Comets."

24. Ibid.; see also Morrison, "Impacts and Evolution."

25. Clarke, "Killer Comets."

26. Associated Press, "After Asteroid Scare, Scientists Agree to Agree," *New York Times*, March 20, 1998, p. A21.

27. O'Connor, "Astronomers Adopt"; see also "Doomsday from 0 to 10," *New York Times*, July 29, 1999, p. A20.

28. "Doomsday from 0 to 10."

29. Reid Detchon, "Climate Change: Moving from Science to Engineering," lecture presented at Board of Direction Meeting, American Society of Civil Engineers, Alexandria, Va., May 2, 2008; Andrew C. Revkin, "At Conference on the Risks to Earth, Few Are Optimistic," *New York Times*, Aug. 24, 2008, p. A11.

CHAPTER 2. ENGINEERING IS ROCKET SCIENCE

1. David Johnson and James Risen, "Nuclear Weapons Engineer Indicted in Removal of Data," *New York Times*, Dec. 11, 1999, pp. A1, A17; David Stout, "Lee's Defenders Say the Scientist Is a Victim of a Witch Hunt Against China," *New York Times*, Dec. 11, 1999, p. A16.

2. Lewis Jordan, *News: How It Is Written and Edited* (New York: New York Times Office of Educational Activities, [1960]), p. 46.

3. See, e.g., Neil A. Lewis, "Imprisoned Scientist Sues U.S. Agencies," *New York Times*, Dec. 21, 1999, p. A28; "Asian-American Scientists File a Bias Complaint," *New York Times*, Dec. 24, 1999, p. A14; Dan Stober and Ian Hoffman, *A Convenient Spy: Wen Ho Lee and the Politics of Nuclear Espionage* (New York: Simon & Schuster, 2001), pp. 20–21.

4. Jordan, *News*, p. 29.

5. Ibid., pp. 29–30; Max Frankel, "Soviet Rocket Hits Moon After 35 Hours; Arrival Is Calculated Within 84 Seconds; Signals Received Till Moment of Impact," *New York Times*, Sept. 14, 1959, pp. 1, 16.

6. See, e.g., Theodore von Kármán, *The Wind and Beyond: Theodore von Kármán, Pioneer in Aviation and Pathfinder in Space* (Boston: Little, Brown, 1967), pp. 211–15.

7. The distinction between scientists and engineers attributed to von Kármán may be found in a variety of forms in numerous secondary sources,

easily found via a search engine on the Web; the author has not been able to identify a primary source.

8. See, e.g., Ronald Breslow, *Chemistry Today and Tomorrow: The Central, Useful, and Creative Science* (Boston: Jones & Bartlett, 1997), p. 93.

9. Thomas P. Hughes, "Einstein the Inventor," *American Heritage of Invention and Technology*, Winter 1991, p. 39.

10. See, e.g., Eugene S. Ferguson, *Engineering and the Mind's Eye* (Cambridge, Mass.: MIT Press, 1992).

11. Compare the choices relating to the design of a motorbike. See, e.g., the cover illustration for Eugene S. Ferguson, "The Mind's Eye: Nonverbal Thought in Technology," *Science* 197 (Aug. 26, 1977): 827–36.

12. Marsha Freeman, *How We Got to the Moon: The Story of the German Space Pioneers* (Washington, D.C.: 21st Century Science Associates, 1993), p. 10.

13. Homer H. Hickam Jr., *Rocket Boys: A Memoir* (New York: Delacorte, 1998), retitled *October Sky* for the paperback edition and movie based on it; Edwin Layton, "Mirror-Image Twins: The Communities of Science and Technology in 19th-Century America," *Technology and Culture* 12, no. 4 (Oct. 1971): 562–80.

14. William Carlos Williams, "A Sort of a Song," in *The Collected Poems of William Carlos Williams*, rev. ed. (New York: New Directions, 1963), p. 7; quoted in Wernher von Braun, "Rocket Propulsion," in Albert Love and James Saxon Childers, eds., *Listen to Leaders in Engineering* (Atlanta and New York: Tupper & Love/David McKay, 1965), p. 119.

15. Jack M. Holl, *Argonne National Laboratory: 1946–1996* (Urbana: University of Illinois Press, 1997).

16. In September 2000, in a Justice Department settlement, Wen Ho Lee pleaded guilty to one count of "mishandling secret information." That same month, a note from the editors of *The New York Times* reflected on the paper's coverage of the case and a long editorial provided a retrospective: "The Times and Wen Ho Lee," *New York Times*, Sept. 26, 2000, p. A2; "The Wen Ho Lee Case," *New York Times*, Sept. 28, 2000, p. A26.

17. Andrew Pollack, "Engineering by Scientists on Embryo Stirs Criticism," *New York Times*, May 13, 2008, p. A14.

18. John Noble Wilford, "A New Breed of Scientists Studying Mars Takes Control," *New York Times*, July 14, 1997, p. A10.

19. Ibid.

20. Guy Gugliotta, "Mapping Celestial Terrains, in All Their 3-D Glory," *New York Times*, Dec. 23, 2008, p. D2.

21. John Dustin Kemper, *The Engineer and His Profession* (New York: Holt, Rinehart & Winston, 1967), p. 1.

22. Ibid.

CHAPTER 3. DOCTORS AND DILBERTS

1. National Academy of Engineering, Committee on Public Understanding of Engineering Messages, *Changing the Conversation: Messages for Improving Public Understanding of Engineering* (Washington, D.C.: National Academies Press, 2008), p. 1.

2. Rose George, *The Big Necessity: The Unmentionable World of Human Waste and Why It Matters* (New York: Metropolitan, 2008); Dwight Garner, "15 Minutes of Fame for Human Waste and Its Never-Ending Assembly Line," *New York Times*, Dec. 12, 2008, p. C34.

3. See, e.g., Henry Petroski, *Invention by Design: How Engineers Get from Thought to Thing* (Cambridge, Mass.: Harvard University Press, 1996), especially Chapter 8 and Figure 8.5. On drug delivery systems, see, e.g., Anne Trafton, "Gold Particles Deliver More than Just Glitter," MIT news release, Dec. 30, 2008, http://www.mit.edu/newsoffice/2008/nanorods-1230.html; Adam Dylewski, "New Drug Delivery System Breaks the 'Mucus Barrier,' " *Medical News Today*, Aug. 22, 2008, http://www.medicalnewstoday.com/articles;118955.php; Chriss Swaney, "Carnegie Mellon Biomedical Engineering Researcher Works to Develop New Drug Delivery System by Using Adult Stem Cells," Carnegie Mellon University Press Release, Nov. 2, 2007, http://www.cit.cmu.edu/media/press/2007/pr_07_Nov02.html.

4. Robin Pogrebin, "Smithsonian Chief Hopes to Institute Big Reforms," *New York Times*, Sept. 15, 2008, pp. E1, E6.

5. C. P. Snow, *The Two Cultures: And a Second Look* (New York: Mentor, 1964), p. 36.

6. Tim Radford, "A Question of Everything," *Guardian*, Aug. 7, 2008, http://www.guardian.co.uk/education/2008/aug/07/research.cern; Dennis Overbye, "Fingers Crossed, Physicists Are Ready for Collider to Roll," *New York Times*, Sept. 9, 2008, p. F2; William Booth, "Smashing Idea," *Washington Post*, Sept. 11, 2008, p. C1.

7. Chris Mooney, *Storm World: Hurricanes, Politics, and the Battle over Global Warming* (Orlando, Fla.: Harcourt, 2007), p. 16; Huxley quoted in Nancy Leveson, "An Introduction to System Safety," (NASA) *ASK Magazine*, Summer 2008, p. 22.

8. Alex R. Dzierba, Curtis A. Meyer, and Eric S. Swanson, "The Search for QCD Exotics," *American Scientist*, Sept.–Oct. 2000, pp. 406–15.

9. On scientific paradigms, see Thomas S. Kuhn, *The Structure of Scientific Revolutions* (Chicago: University of Chicago Press, 1962).

10. Chris Impey, *The Living Cosmos: Our Search for Life in the Universe* (New York: Random House, 2007), p. 9; Cornelia Dean, "Handle with Care," *New York Times*, Aug. 12, 2008, pp. F1, F4.

11. But see Steen Hyldgaard Christensen, Martin Meganck, and Bernard Delahousse, *Philosophy in Engineering* (Arhus, Denmark: Academica, 2007).

12. Natalie Angier, *The Canon: A Whirligig Tour of the Beautiful Basics of Science* (Boston: Houghton Mifflin, 2007), pp. 2–3.

13. Ibid., p. 14.

14. C. C. Furnas, Joe McCarthy, and the Editors of *Life*, *The Engineer* (New York: Time Inc., 1966), p. 9.

15. See http://dilbert.com/strips, and especially strips for Jan. 14, Jan. 22, Jan. 23, Jan. 24, and Feb. 14, 2009.

16. See, e.g., http://www.youtube.com/watch?v=CmYDgncMhXw.

17. Steve Lohr, "The Rise of the Humble Engineer," *New York Times*, June 17, 2008, pp. C1, C4.

18. Ibid., p. C1.

19. Ibid., p. C4.

CHAPTER 4. WHICH COMES FIRST?

1. Barry Allen, *Artifice and Design: Art and Technology in Human Experience* (Ithaca, N.Y.: Cornell University Press, 2008), p. 117.

2. Norbert J. Delatte, Jr., *Beyond Failure: Forensic Case Studies for Civil Engineers* (Reston, Va.: ASCE Press, 2009), p. 153.

3. C. P. Snow, *The Two Cultures: And a Second Look* (New York: Mentor, 1964), p. 36; James Randerson, " 'Giant Microscope' That Peers into the Heart of a Structure," *Guardian*, May 15, 2008, http://www.guardian.co.uk; for more on the physicist's point of view, see, e.g., Irwin Goodwin, "Engineers Proclaim Top Achievements of 20th Century, but Neglect Attributing Feats to Roots in Physics," *Physics Today*, May 2000, pp. 48–49.

4. Randerson, " 'Giant Microscope.' "

5. Andrew Pollack, "A Biotech Superstar Looks at the Bigger Picture," *New York Times*, April 17, 2001, p. F3.

6. Bruce Alberts, "Celebrating a Year of *Science*," *Science*, Dec. 19, 2008, p. 1757.

7. Jules Verne, *De la Terre à la Lune*, 1865; Arthur C. Clarke, "V2 for Ionosphere Research?," *Wireless World*, Feb. 1945, p. 58; Clarke quoted in Laurence Prusak, "Authors Who Make a Difference," (NASA) *ASK Magazine*, Spring 2008, p. 54; Gerald Jonas, "Arthur C. Clarke, Author Who Saw Science Fiction Become Real, Dies at 90," *New York Times*, March 19, 2008, national edition, p. C12.

8. L. T. C. Rolt, *Isambard Kingdom Brunel* (Hammondsworth, Middlesex: Penguin, 1957), pp. 249–60; Curt Wohleber, "The Annihilation of

Time and Space," *American Heritage of Invention and Technology*, Spring/ Summer 1991, http://www.americanheritage.com/articles/magazine/it/ 1991/1/1991_1_20.shtml.

9. See, e.g., Henry Petroski, "Harnessing Steam," in *Remaking the World: Adventures in Engineering* (New York: Alfred A. Knopf, 1997), pp. 117–25.

10. Gabrielle Walker, *An Ocean of Air: Why the Wind Blows and Other Mysteries of the Atmosphere* (Orlando, Fla.: Harcourt, 2007), pp. 161, 169, 174–75, 188.

11. See, e.g., Walter G. Vincenti, *What Engineers Know and How They Know It: Analytical Studies from Aeronautical History* (Baltimore: Johns Hopkins University Press, 1990); Tom Crouch, *The Bishop's Boys: A Life of Wilbur and Orville Wright* (New York: W. W. Norton, 1989), pp. 175, 242–43, 395.

12. Orville Wright, *How We Invented the Airplane: An Illustrated History*, ed. Fred C. Kelly (New York: Dover, 1988), pp. v, 5, 12.

13. Simon Ramo, "The Design of the Whole—Systems Engineering," in Albert Love and James Saxon Childers, eds., *Listen to Leaders in Engineering* (Atlanta and New York: Tupper & Love/David McKay, 1965), p. 164.

14. Lord Kennet, quoted in Richard Elliot Benedick, *Ozone Diplomacy: New Directions in Safeguarding the Planet*, enlarged ed. (Cambridge, Mass.: Harvard University Press, 1998), p. 2.

15. Allen, *Artifice and Design*, p. 128.

16. Ibid., p. 130.

17. Gavin Weightman, *The Industrial Revolutionaries: The Creation of the Modern World, 1776–1914* (New York: Atlantic Monthly Press, 2009), p. 310; Elizabeth Kolbert, "Dymaxion Man," *New Yorker*, June 9 & 16, 2008, pp. 64–69; see also James Sterngold, "The Love Song of R. Buckminster Fuller," *New York Times*, June 15, 2008, p. AR26.

18. Neil Schlager, ed., *When Technology Fails: Significant Technological Disasters, Accidents, and Failures of the Twentieth Century* (Detroit: Gale Research, 1994), pp. 591–97.

19. Allen, *Artifice and Design*, p. 120.

20. Sevcik and Wetzel, "Engineers Not Scientists Will Rescue the Economy," post on App Performance Review blog, Feb. 10, 2009, http:// www.networkworld.com/community/node/38419.

CHAPTER 5. EINSTEIN THE INVENTOR

1. On the National Inventors Council, see http://www.ipmall.info/hosted _resources/nic.asp.

2. The Lamarr-Anthiel story is the basis for the play *Frequency Hopping*, by Elyse Singer; "Mechanical Dreams Come True," by Anthony Tommasini, *New York Times*, June 9, 2008, pp. E1, E5; "Female Inventors: Hedy Lamarr," http://www.inventions.org/culture/female/lamarr .html; "Hedy Lamarr Inventor," *New York Times*, Oct. 1, 1941, p. 24; Hedy Kiesler Markey and George Antheil, "Secret Communication System," U.S. Patent No. 2,292,387 (Aug. 11, 1942).

3. Http://www.paulwinchell.com; Adam Bernstein, "TV Ventriloquist, Cartoon Voice and Inventor Paul Winchell Dies," *Washington Post*, June 27, 2005, p. B4; Julie Salamon, "Paul Winchell, 82, TV Host and Film Voice of Pooh's Tigger, Dies," *New York Times*, June 27, 2005; Paul Winchell, "Artificial Heart," U.S. Patent No. 3,097,366 (July 16, 1963); see also Sandra Blakeslee, "Willem Kolff, Doctor Who Invented Kidney and Heart Machines, Dies at 97," *New York Times*, Feb. 13, 2009, p. A23.

4. Frederick Emmons Terman, "Engineering Research," in Albert Love and James Saxon Childers, eds., *Listen to Leaders in Engineering* (Atlanta and New York: Tupper & Love/David McKay, 1965), p. 39.

5. National Academy of Engineering, *Directory of Members and Foreign Associates* (Washington, D.C.: National Academy of Engineering, 2006), pp. 320–21. Of the almost 2,100 NAE members (active and emeritus) as of mid-2006, over 150 (or more than 7 percent) were members of both the National Academy of Engineering and the National Academy of Sciences. (Twenty-six NAE members were also members of the Institute of Medicine, and six were members of all three national academies.) On Wozniak, see Ashlee Vance, "Wozniak Accepts Post at a Storage Start-up," *New York Times*, Feb. 4, 2009, p. B3.

6. Ad Maas, "Einstein as Engineer: The Case of the Little Machine," *Physics in Perspective* 9 (2007): 305–28.

7. Thomas P. Hughes, "Einstein, Inventors, and Invention," *Science in Context* 6, no. 1 (1993): 25–42; Maas, "Einstein as Engineer," p. 306; Walter Isaacson, *Einstein: His Life and Universe* (New York: Simon & Schuster, 2007), pp. 22–23.

8. Isaacson, *Einstein*, pp. 24, 32. Zurich Polytechnic developed into the distinguished Eidgenössische Technische Hochschule, familiarly known as ETH.

9. Thomas P. Hughes, "Einstein the Inventor," *American Heritage of Invention and Technology*, Winter 1991, pp. 34–39.

10. Ibid., p. 39.

11. Matthew Trainer, "Albert Einstein's Patents," *World Patent Information* 28 (2006): 159–65.

12. See http://www.dannen.com/chronbio.html.

13. Gene Dannen, "The Einstein-Szilárd Refrigerators," *Scientific American*, January 1997, pp. 90–95.

14. Abraham Pais, *Subtle Is the Lord: The Science and the Life of Albert Einstein* (Oxford: Oxford University Press, 2005), p. 490.

15. Dannen, "The Einstein-Szilárd Refrigerators."

16. Andrew D. Delano, "Design Analysis of the Einstein Refrigeration Cycle," Ph.D. thesis, School of Mechanical Engineering, Georgia Institute of Technology, 1998. See also http://www.me.gatech.edu/energy/andy; Georg Alefeld, "Einstein as Inventor," *Physics Today*, May 1980, pp. 9, 11, 13.

17. For the Einstein-Szilárd refrigeration system, the basic British Patent No. 282,428, issued in 1928, is essentially the same as Albert Einstein and Leo Szilárd, "Refrigeration," U.S. Patent No. 1,781,541, issued on Nov. 11, 1930.

18. Dannen, "The Einstein-Szilard Refrigerators"; see also Steve Silverman, *Einstein's Refrigerator and Other Stories from the Flip Side of History* (Kansas City, Mo.: Andrews McMeel, 2001), pp. 60–63.

19. David C. McHenry Jr., "Curly Cord Automatic Binding Tie," U.S. Patent No. 5,515,580 (May 14, 1996).

CHAPTER 6. SPEED BUMPS

1. Eoin O'Carroll, "EU Bans Incandescent Light Bulbs," *Christian Science Monitor*, Oct. 15, 2008; John M. Broder, "Obama Orders New Rules to Raise Energy Efficiency," *New York Times*, Feb. 6, 2009, p. A14.

2. Michael Kanellos, "Father of the Compact Fluorescent Bulb Looks Back," *ZDNet News*, Aug. 16, 2007, http://news.zdnet.com/2100-9595_22-160128.html; "Edward E. Hammer, 1931– ," *IEEE History Center*, http://www.ieee.org/web/aboutus/history_center/biography/hammer.html; Stephanie Rosenbloom, "Home Depot Offers Recycling for Compact Fluorescent Bulbs," *New York Times*, June 24, 2008, pp. C1, C7.

3. "A Cloth to Cut the Mercury Risk from Light Bulbs," *New York Times*, July 8, 2008, p. F3; Richard Lewis, "Brown Researchers Create Mercury-Absorbent Container Linings for Broken CFLs," Brown University news release, June 27, 2008, http://www.news.brown.edu/pressreleases/2008/06/mercurycloth08jun.

4. Sari Krieger, "Bright Future," *Wall Street Journal*, Sept. 15, 2008, p. R4; Eric A. Taub, "Philips Sees Light, and Patients, at the End of Its Overhaul," *New York Times*, Dec. 26, 2008, p. B4.

5. Eric A. Taub, "Would You Buy These Weird Bulbs?" *New York Times*, Nov. 10, 2008, p. B8.

6. Traffic Logix, "Speed Humps," http://www.trafficlogix.com/speed-

humps.asp; National Motorists Association, "Traffic Calming," http://www.motorists.org/trafficcalm; Word Spy, "Sleeping policeman," http://www.wordspy.com/words/sleepingpoliceman.asp.

7. National Motorists Association, "Traffic Calming"; for a view on how traffic signs can make the road less safe, see John Staddon, "Distracting Miss Daisy," *Atlantic*, July/Aug. 2008, http://www.theatlantic.com/doc/200807/traffic.

8. William F. Baker, "Uneventful Is Not Easy," letter to the editor, *Engineering News-Record*, June 9, 2008, p. 7; Rebecca Mead, "Winged Victories," *New Yorker*, Sept. 1, 2008, p. 108.

9. "The New York Times Building," http://events.nydailynews.com/new-york-ny/venues/show/866752-the-new-york-times-building.

10. Paul Goldberger, "Dream House," *New Yorker*, Oct. 30, 2000, p. 108.

11. Http://newyorktimesbuilding.com; see also Jeffrey A. Callow, Kyle E. Krall, and Thomas Z. Scarangello, "Ahead of the Times," *Civil Engineering*, Nov. 2008, pp. 54–63; see also Jeffrey A. Callow, Kyle E. Krall, and Thomas Z. Scarangello, "Inside Out," *Modern Steel Construction*, Jan. 2009, pp. 21–25.

12. Kareem Fahim, "The People Who Aspire to Great Heights, Literally, Hand over Hand," *New York Times*, June 7, 2008, p. B2; James Barron and Robin Pogrebin, "Surprised by Climbers' Stunts, Times Adds Security Measures," *New York Times*, June 7, 2008, pp. B1–B2.

13. Barron and Pogrebin, "Surprised by Climbers' Stunts."

14. James Barron, "2 Men Scale New York Times Building Hours Apart," *New York Times*, June 6, 2008, pp. B1, B6; Jim Dwyer, "The Explanation: Climb Here to Combat Malaria," *New York Times*, June 11, 2008, pp. B1, B4.

15. John Eligon, "Most Charges Dropped Against Times Climber," *New York Times*, June 13, 2008, p. B2; see also John Eligon, "Times Building Climber Is Sentenced to Three Days of Community Service," *New York Times*, Dec. 2, 2008, p. A31.

16. John Eligon, "Climbers of Times Building Get Different Jury Treatment," *New York Times*, Oct. 8, 2008, p. A26; http://thesolutionissimple.org. See also http://www.alainrobert.com; "Second Times Climber Pleads Guilty," *New York Times*, Dec. 20, 2008, p. A22.

17. Michael Barbaro and John Eligon, "City Seeks Bigger Penalty for Those Who See Skyline as a Jungle Gym," *New York Times*, June 27, 2008, pp. B1, B7; Sewell Chan, "After 3rd Climber, Times Alters Its Building's Facade," *New York Times*, July 10, 2008, p. B2.

18. Chan, "After 3rd Climber"; Barron and Pogrebin, "Surprised by Climbers' Stunts"; Sewell Chan, "Architect Supports Changes to Times Tower," *New York Times*, July 11, 2008, p. B3.

19. Barron and Pogrebin, "Surprised by Climbers' Stunts"; Chan, "Architect Supports Changes."
20. Http://www.kongisking.net/history; A. O. Scott, "Walking on Air Between the Towers," *New York Times*, July 25, 2008, pp. C1, C16; Fahim, "The People Who Aspire to Great Heights."
21. Gavin Weightman, *The Industrial Revolutionaries: The Creation of the Modern World, 1776–1914* (New York: Atlantic Monthly Press, 2009), pp. 300–301.
22. Michael H. Westbrook, *The Electric Car: Development and Future of Battery, Hybrid and Fuel-Cell Cars* (London: Institution of Electrical Engineers, 2001), pp. 15–16; Daniel Gross et al., *Forbes Greatest Business Stories of All Time* (New York: Wiley, 1996), http://www.wiley.com/legacy/products/subject/business/forbes/ford.html.
23. Westbrook, *The Electric Car*, pp. 16–19.
24. Alex Kuczynski, "Don't Know What It Is, or What It's About, but Harvard Thinks It's Worth $250,000," *New York Times*, Jan. 12, 2001, p. C6.
25. Steve Kemper, *Code Name Ginger: The Story Behind Segway and Dean Kamen's Quest to Invent a New World* (Boston: Harvard Business School Press, 2003), pp. 117, 302–3; Eric A. Taub, "Drawing Stares and the Police but Not a Whole Lot of Buyers," *New York Times*, Aug. 9, 2003, pp. C1, C2.
26. Rachel Metz, "Oft-Scorned Segway Finds Friends Among the Disabled," *New York Times*, Oct. 14, 2004, p. G5; Kemper, *Code Name Ginger*, pp. 154, 169, 304; John Schwartz, "Scooter Much Publicized for Stability Is Recalled," *New York Times*, Sept. 27, 2003, p. A8.
27. Schwartz, "Scooter Much Publicized"; Sara Ivry, "Segway Plans 2 New Models of Transporter," *New York Times*, Aug. 14, 2006, p. C6.
28. Ivry, "Segway Plans"; Thomas J. Lueck, "Taller, Faster Officers on Patrol as Police Dept. Gets 10 Segways," *New York Times*, May 17, 2007, p. B1; April Dembosky, "Outlaws on 2 (Battery-Powered) Wheels," *New York Times*, Aug. 12, 2008, p. B6.
29. Matt Richtel, "Lost in E-Mail, Tech Firms Face Self-Made Beast," *New York Times*, June 14, 2008, pp. A1, A14.
30. Ibid.
31. Ibid.
32. "ABC and Writers Skirmish over After-Hours E-Mail," *New York Times*, June 23, 2008, p. C4.
33. Nicholas Carr, "Is Google Making Us Stupid?" *Atlantic*, July/Aug. 2008, http://www.theatlantic.com/doc/200807/google.
34. Damon Darlin, "Technology Doesn't Dumb Us Down. It Frees Our Minds," *New York Times*, Sept. 21, 2008, http://www.nytimes.com/2008/09/21/technology/21ping.html.

CHAPTER 7. RESEARCH AND DEVELOPMENT

1. Ronald R. Kline and Thomas C. Lassman, "Competing Research Traditions in American Industry: Uncertain Alliances Between Engineering and Science at Westinghouse Electric, 1886–1935," *Enterprise & Society* 6, no. 4 (2005): 601–45; Matthew Josephson, *Edison: A Biography* (New York: Wiley, 1992), p. 314, quoted at http://www.nps.gov/history/nr/twhp/wwwlps/lessons/25edison/25edison.htm; Paul Israel, *Edison: A Life of Invention* (New York: Wiley, 1998), p. 177.

2. Israel, *Edison*, pp. 321, 336; Kline and Lassman, "Competing Research Traditions," pp. 602–4.

3. Kline and Lassman, "Competing Research Traditions," pp. 604–5.

4. Much of this paragraph is abridged from the essay on Steinmetz in Henry Petroski, *Remaking the World: Adventures in Engineering* (New York: Alfred A. Knopf, 1997), pp. 3–4; see also "The GE Global Research Story," http://www.ge.com/research/grc_8_1.html.

5. Petroski, *Remaking the World*, pp. 7–8.

6. All typefaces referred to and used in this paragraph are in Microsoft Word. On the ampersand, see Michael Quinion, "World Wide Words: Ampersand," http://www.worldwidewords.org/weirdwords/ww-amp1.htm.

7. Thomas P. Hughes, *Rescuing Prometheus* (New York: Vintage, 2000), pp. 154–55; see also http://www.rand.org/about.

8. Kline and Lassman, "Competing Research Traditions," pp. 605–6.

9. Ibid., pp. 606–7.

10. Ibid., pp. 611–13, 615; Skinner quoted ibid., p. 613.

11. Ibid., pp. 617–18.

12. Ibid., pp. 618–19.

13. Ibid., pp. 629, 631.

14. Margaret B. W. Graham and Alec T. Shuldiner, *Corning and the Craft of Innovation* (New York: Oxford University Press, 2001), pp. 21–22, 34–35.

15. Ibid., pp. 41–42, 44.

16. Ibid., pp. 43, 52.

17. Ibid., pp. 14, 55–58, 60–61; Kenneth Chang, "Anything but Clear," *New York Times*, July 29, 2008, pp. F1, F4.

18. Graham and Shuldiner, *Corning and the Craft*, pp. 62–65.

19. Ibid., p. 66; chemist quoted ibid., p. 75.

20. Kline and Lassman, "Competing Research Traditions," p. 639.

CHAPTER 8. DEVELOPMENT AND RESEARCH

1. Margaret B. W. Graham and Alec T. Shuldiner, *Corning and the Craft of Innovation* (New York: Oxford University Press, 2001), p. 14; Einstein's letter to Roosevelt quoted from Walter Isaacson, *Einstein: His Life and Universe* (New York: Simon & Schuster, 2007), pp. 22, 474.

2. Vannevar Bush, "As We May Think," *Atlantic*, July 1945, http://www.theatlantic.com/doc/194507/bush.

3. Daniel J. Kevles, *The Physicists: The History of a Scientific Community in Modern America* (New York: Alfred A. Knopf, 1977), pp. 293–96; see also Stanley Goldberg, "Inventing a Climate of Opinion: Vannevar Bush and the Decision to Build the Bomb," *Isis* 83, no. 3 (Sept. 1992): 429–30; "Internet Pioneers: Vannevar Bush," http://www.ibiblio.org/pioneers/bush.html; "Yankee Scientist," *Time*, April 3, 1944, http://www.time.com/time/magazine/article/0,9171,850430-1,00.html.

4. Kevles, *The Physicists*, pp. 296–97; Goldberg, "Inventing a Climate," pp. 430–31; "Yankee Scientist," *Time*, April 3, 1944, cover story.

5. Michael Eckert and Helmut Schubert, *Crystals, Electrons, Transistors: From a Scholar's Study to Industrial Research*, Thomas Hughes, trans. (New York: Springer, 1997), p. 140; Kevles, *The Physicists*, pp. 300, 342; *Time*, April 3, 1944, front cover.

6. Kevles, *The Physicists*, pp. 343–47.

7. Vannevar Bush, *Science—the Endless Frontier* (Washington, D.C.: U.S. Government Printing Office, 1945), p. 2. A full (but unpaginated) text of Bush's "Report to the President" is available at http://www.nsf.gov/about/history/vbush1945.htm.

8. Kevles, *The Physicists*, pp. 349–59.

9. See, e.g., Galileo, *Dialogue Concerning the Two Chief World Systems*, Stillman Drake, trans., 2nd rev. ed. (Berkeley: University of California Press, 1992); Galileo Galilei, *Dialogues Concerning Two New Sciences*, Henry Crew and Alfonso de Salvio, trans., reprint (New York: Dover, [1954]).

10. Galileo, *Two New Sciences*, pp. 1–6.

11. Neal Lane, "US Science and Technology: An Uncoordinated System that Seems to Work," *Technology in Society* 30 (2008): 252; Jay Holmes, ed., *Energy, Environment, Productivity. Proceedings of the First Symposium on RANN: Research Applied to National Needs* (Washington, D.C.: U.S. Government Printing Office, 1974).

12. Ronald R. Kline and Thomas C. Lassman, "Competing Research Traditions in American Industry: Uncertain Alliances Between Engineering and Science at Westinghouse Electric, 1886–1935," *Enterprise & Society* 6, no. 4 (2005): 639.

13. Chalmers W. Sherwin and Raymond S. Isenson, "Project Hindsight," *Science* 156 (June 23, 1967): 1571–77; quotations are from Edwin Layton, "Mirror-Image Twins: The Communities of Science and Technology in 19th-Century America," *Technology and Culture* 12, no. 4 (Oct. 1971): 563–64; Lane, "US Science and Technology," p. 250; Adil Ahmed, "Origins of Dated Federal R&D Policy," *Science Progress*, July 24, 2008, http://www.scienceprogress.org/2008/07/origins-of-outdated-federal-rd-policy.

14. Vannevar Bush, "The Engineer," in Albert Love and James Saxon Childers, eds., *Listen to Leaders in Engineering* (Atlanta and New York: Tupper & Love/David McKay, 1965), pp. 9–12; see also Sunny Y. Auyang, *Engineering—an Endless Frontier* (Cambridge, Mass.: Harvard University Press, 2004), p. 3.

15. Bush, "The Engineer," pp. 7–9.

16. Ibid., p. 11; see also Paul A. Heckert, "Karl Guthe Jansky: Brief Biography of the Father of Radio Astronomy," *Suite101.com*, http://greatscientists.suite101.com/article.cfm/karl_guthe_jansky; Harald T. Friis, "Karl Jansky: His Career at Bell Telephone Laboratories," *Science* 149, no. 3686 (Aug. 20, 1965): 841–42; J. E. Brittain, "Jansky Discovers Extraterrestrial Radio Noise," *Proc. IEEE* 72 (June 1984): 709; Bernard M. Oliver, "Electronics," in Love and Childers, eds., *Listen to Leaders*, pp. 225–26.

17. Graham and Shuldiner, *Corning and the Craft*, pp. 370–71.

18. Ahmed, "Origins of Dated Federal Policy"; Lane, "US Science and Technology," p. 252, 259n29.

19. Graham and Shuldiner, *Corning and the Craft*, p. 138; Corning, Inc., *Annual Report*, 2007, p. 19.

20. [National Research Council], *Allocating Federal Funds for Science and Technology* (Washington, D.C.: National Academies Press, 1995).

21. Bush, *Science*, pp. 18–21.

22. George E. Holbrook, "Chemical Engineering," in Love and Childers, eds., *Listen to Leaders*, p. 67.

23. Graham and Shuldiner, *Corning and the Craft*, pp. 24–25.

24. Barry Allen, *Artifice and Design: Art and Technology in Human Experience* (Ithaca, N.Y.: Cornell University Press, 2008), p. 115.

25. Chris Bryant, "US Retains Top Spot in World Technology League," *FT.com*, June 12, 2008; Ahmed, "Origins of Dated Federal Policy."

26. National Academies, *Rising Above the Gathering Storm: Energizing and Employing America for a Brighter Economic Future* (Washington, D.C.: National Academies Press, 2006), p. 1; Bill Buxton, "The Price of Forgoing Basic Research," *BusinessWeek.com*, Dec. 18, 2008; Claire Cain Miller, "Another Voice Warns of an Innovation Slowdown," *New York*

Times, Sept. 1, 2008, p. C4; Corning, *Annual Report*, 2007, p. 19; Katie Hafner, "Microsoft Adds Research Lab in East as Others Cut Back," *New York Times*, Feb. 4, 2008, p. C3.

27. American Association for the Advancement of Science, "Final Stimulus Bill Provides $21.5 Billion for Federal R&D," Feb. 16, 2009, update, http://www.aaas.org/spp/rd/stim09c.htm.

28. Geoffrey Carr, "The Power and the Glory," *Economist*, June 19, 2008, http://www.economist.com/opinion/displaystory.cfm?story_id=11565685.

29. Http://www.copenhagenconsensus.com; Bjørn Lomborg, "How to Get the Biggest Bang for 10 Billion Bucks," *Wall Street Journal*, July 28, 2008, p. A15.

CHAPTER 9. ALTERNATIVE ENERGIES

1. *Bartlett's Familiar Quotations*, 16th ed., Justin Kaplan, gen. ed. (Boston: Little, Brown, 1992), p. 715; Oppenheimer is quoting from the *Bhagavad Gita*.

2. Philip Sporn, "Energy and Energy Conversion," in Albert Love and James Saxon Childers, eds., *Listen to Leaders in Engineering* (Atlanta and New York: Tupper & Love/David McKay, 1965), p. 199.

3. Nicholas Confessore, "In Rural New York, Windmills Can Bring Whiff of Corruption," *New York Times*, Aug. 18, 2008, pp. A1, A16.

4. Associated Press, "N.J. Vows 'Race to the Sea' for Wind Power," Oct. 7, 2008, http://www.msnbc.msn.com/id/27068747; Ken Belson, "New Jersey Grants Rights to Build a Wind Farm About 20 Miles Offshore," *New York Times*, Oct. 4, 2008, p. B2; Kari Lydersen, "Studies Lift Hope for Great Lakes Wind Turbine Farms," *Washington Post*, Oct. 7, 2008, p. A9.

5. Lydersen, "Studies Lift Hope."

6. Michael Barbaro, "Bloomberg Offers Windmill Power Plan," *New York Times*, Aug. 20, 2008, pp. B1, B6; Ken Belson and David W. Dunlap, "Architects and Engineers Express Doubt About Bloomberg's Windmill Proposal," *New York Times*, Aug. 21, 2008, p. B6; Kate Galbraith, "Personal-Size Wind Power," *New York Times*, Sept. 4, 2008, p. C9.

7. Felicity Barringer, "Indian Tribes See Profit in Harnessing the Wind for Power," *New York Times*, Oct. 10, 2008, p. A17; Tom Wright, "Winds Shift for Renewable Energy as Oil Price Sinks, Money Gets Tight," *Wall Street Journal*, Oct. 20, 2008, *WSJ.com*, http://online.wsj.com/article/SB122446199550848849.html; Clifford Krauss, "Alternative Energy Suddenly Faces Headwinds," *New York Times*, Oct. 21, 2008, pp. B1, B10.

8. Alexis Madrigal, "DOE Report: Wind Could Power 20 Percent of US

Grid by 2030," http://www.wired.com/wiredscience/2008/05/doe-report-wired; U.S. Department of Energy, "20% Wind Energy by 2030: Increasing Wind Energy's Contribution to U.S. Electricity Supply," May 2008; see http://20percentwind.org/20percent_wind_energy _report_05-11-08_wk.pdf.

9. Galbraith, "Personal-Size Wind Power," pp. C1, C9; "Cascade Engineering Launches Sales of Small Wind Turbine," *WWJ* (Southfield, Mich.) *Newsradio 950*, Oct. 27, 2008.

10. Mark Clayton, "How White Roofs Shine Bright Green," *Christian Science Monitor*, Oct. 3, 2008; http://www.csmonitor.com/centennial/events/monitors-stories/2008/09/how-white-roofs-shine-bright-green.

11. Alan S. Brown, "Solar Cells at $1 a Watt?" *Mechanical Engineering*, May 2008, p. 72; Alan S. Brown, "The Emerging Alternate Energy Consensus," *The Bent of Tau Beta Pi*, Fall 2008, p. 20; Tara Siegel Bernard and Kate Galbraith, "Bailout Brings with It Diverse Perks," *New York Times*, Oct. 4, 2008, p. C6.

12. Kate Galbraith, "Solar Panels Are Vanishing, Only to Reappear on the Internet," *New York Times*, Sept. 24, 2008, pp. C1, C15.

13. Mark Landler, "Solar Valley Rises in an Overcast Land," *New York Times*, May 16, 2008, p. C1.

14. Nicholas Kulish, "German City Wonders How Green Is Too Green," *New York Times*, Aug. 7, 2008, p. A8.

15. Dan Frosch, "Citing Need for Assessments, U.S. Freezes Solar Energy Projects," *New York Times*, June 27, 2008, p. A13; Peter Maloney, "Solar Projects Draw New Opposition," *New York Times*, Sept. 24, 2008, "Business of Green" special section, p. 2.

16. Barringer, "Indian Tribes See Profit," p. A17; Matthew L. Wald, "Two Large Solar Power Plants Are Planned for California," *New York Times*, Aug. 15, 2008, p. C4.

17. Stephanie Tavares, "Solar Takes No Shine to Nevada," *Las Vegas Sun*, Oct. 26, 2008.

18. Anthony DePalma, "New Jersey Dealing with Solar Policy's Success," *New York Times*, June 25, 2008, pp. B1, B7; Tudor Van Hampton and Thomas F. Armistead, "Clean Power Play," *Engineering News-Record*, June 30/July 7, 2008, pp. 22–27.

19. James Kanter, "The Trouble with Markets for Carbon," *New York Times*, June 20, 2008, pp. C1, C5.

20. Brown, "The Emerging Alternate Energy Consensus," pp. 21–22; Blaine Harden, "Filipinos Draw Power from Buried Heat," *Washington Post*, Oct. 4, 2008, p. A1; Sarah Lyall, "Smokestacks in a White Wilderness Divide Iceland in a Development Debate," *New York Times*, Feb. 4, 2007, pp. 1, 16.

21. Charles J. Murray, "It's Not a Slam Dunk," *Design News*, April 28, 2008, p. 38.

22. Nick Bunkley, "Plug-in Hybrid from G.M. Is Nearly Ready for Testing," *New York Times*, Aug. 15, 2008, p. C9; General Motors, "Our Challenge: Nothing Short of Reinventing the Automobile," display ad, *New York Times*, Aug. 12, 2008, p. A7; Bill Vlasic, "Chrysler Enters the Race to Introduce Electric Models," *New York Times*, Sept. 24, 2008, p. C4; Bernard and Galbraith, "Bailout Brings."

23. Matt Nauman and Pete Carey, "Tesla Motors Announces Layoffs; Musk to Assume CEO Post," (San Jose, Calif.) *Mercury News*, Oct. 15, 2008, http://www.mercurynews.com/cars/ci_10727401?nclick_check=1; Claire Cain Miller, "Tesla Says It Will Lay Off Employees and Delay Its All-Electric Sedan Until 2011," *New York Times*, Oct. 16, 2008, p. B4; Charles J. Murray, "Auto Industry Working Hard to Make an Electric Vehicle Battery," *Design News*, April 14, 2008, quoted in American Society of Civil Engineers, *ASCE SmartBrief*, April 15, 2008; http://wuphys.wustl.edu/~katz/naturefooled.html; Randall Stross, "Only the Rich Can Afford It. Should Taxpayers Back It?" *New York Times*, Nov. 30, 2008, p. BU3.

24. Isabel Ordóñez, "Everybody into the Ocean," *Wall Street Journal*, Oct. 6, 2008, p. R6; Jasper Copping, "Ocean Currents Can Power the World, Say Scientists," *Telegraph.com.uk*, Nov. 29, 2008; Canadian Press, "Tidal Power Projects Move Forward in Eastern Maine and Around the World," *ENR.com*, "Digital Wire," Sept. 7, 2008.

25. Brian Skoloff, "Scientists Hoping to Tap into Ocean's Power," (Durham, N.C.) *Herald-Sun*, Feb. 1, 2008; "Harnessing the Power of the Gulf Stream," *NPR Morning Edition*, Dec. 3, 2007, p. 1.

26. Skoloff, "Scientists Hoping"; Robin Shulman, "N.Y. Tests Turbines to Produce Power," *Washington Post*, Sept. 20, 2008, p. A2.

27. Terry Kinney and Jim Suhr, "Mandates Driving Surge to the River for Hydropower," Associated Press, Nov. 30, 2008; Copping, "Ocean Currents"; Bill Bartleman, "Interior Hopes to Stall River Turbines: Number of Permits Draws Ire of Agency Made Up of PPS, Princeton Company," Associated Press, Nov. 30, 2008.

28. Chris Gourlay, "Light Fantastic: Pedestrians to Generate Power," (London) *Sunday Times*, June 8, 2008; p. 7; Tom Abate, "Latest in Battle Gear Is a Prize: Batteries," *San Francisco Chronicle*, June 22, 2008, p. C-1.

29. Elisabeth Rosenthal, "Partying Helps Power a Dutch Nightclub That Harnesses the Energy of Youth," *New York Times*, Oct. 24, 2008, p. A8.

30. Bob Keefe, "Tiny Wires, Our Motion Could Power Devices," *Columbus* (Ohio) *Dispatch*, Nov. 17, 2008, http://www.dispatch.com/live/

content / business / stories / 2008/11/17/small_power_-_cox.ART_ART
_11-17-08_C11_9ABT8AV.html.

31. Rick Perry, "Texas Is Fed Up with Corn Ethanol," *Wall Street Journal*, Aug. 12, 2008, p. A21; "Stop the Rush to Corn Biofuel," editorial, *Christian Science Monitor*, Aug. 12, 2008, p. 8.

32. Alexei Barrionuevo, "Ethanol Could Corrode Pumps, Testers Say," *New York Times*, Oct. 27, 2006, p. C10.

33. Richard Morgan, "Beyond Carbon: Scientists Worry About Nitrogen's Effects," *New York Times*, Sept. 2, 2008, p. F3; Andrew C. Revkin, "Study Proposes New Strategy to Stem Global Warming," *New York Times*, Aug. 19, 2008, p. A13.

34. Experts quoted from Jad Mouawad, "Promise of Biofuel Clouded by Weather Risks," *New York Times*, July 1, 2008, pp. A1, A17.

35. Elisabeth Rosenthal, "Once a Dream Fuel, Palm Oil May Be an Eco-Nightmare," *New York Times*, Jan. 31, 2007, p. C1.

36. Simon Ramo, "The Design of the Whole—Systems Engineering," in Love and Childers, eds., *Listen to Leaders*, p. 167; Cross quoted in James R. Killian Jr., "Engineering Obligations in Government and Public Affairs," in Love and Childers, eds., *Listen to Leaders*, pp. 313–14.

37. Robert Gavin, "Recipes for New Fuels Reviving Maine's Mills," *Boston Globe*, Aug. 18, 2008, http://www.boston.com/news/local/articles/2008/08/18/recipes_for_new_fuels_reviving_maines_mills; Associated Press, "Company Seeks $7M for Algae Fuel Test in Holland," June 22, 2008.

38. Alexei Barrionuevo, "Rise in Ethanol Raises Concerns About Corn as a Food," *New York Times*, Jan. 5, 2007, p. C7; Kate Galbraith, "Economy Shifts, and the Ethanol Industry Reels," *New York Times*, Nov. 5, 2008, pp. B1, B7.

39. Jim Romeo, "Inventor's Fuel-Saving Mud Flaps Make Splash with Truckers," *Engineering News-Record*, June 16, 2008, p. 59; Gail Collins, "The Energy Drill," *New York Times*, Aug. 7, 2008, p. A25.

40. Verlyn Klinkenborg, "Some Doubts upon Entering a New Carboniferous Era," *New York Times*, June 24, 2008, p. A22.

41. William M. Bulkeley, "The Search for a Better Battery Seems Everlasting," *Wall Street Journal*, Oct. 28, 2008, p. A11; Associated Press, "Fuel-Cell Powered Devices Getting Closer," *The Motley Fool*, Dec. 1, 2008, http://www.fool.com/news/associated-press/2008/12/01/fuel-cell-powered-devices-getting-closer.aspx.

42. David Biello, "Inside the Solar-Hydrogen House: No More Power Bills—Ever," *Scientific American*, June 19, 2008, http://www.sciam.com/article.cfm?id=hydrogen-house.

43. Duncan Haimerl, "Showcase Tour of Hydrogen-Fueled Cars Stops in State," *Hartford* (Conn.) *Courant*, Aug. 13, 2008, http://www

.courant.com/business/hc-hydrogen0813; Erin Ailworth, "State's First Hydrogen Refueling Facility Powers Up," *Boston Globe*, Aug. 12, 2008, http://www.boston.com/business/articles/2008/08/12/states_first_hydrogen_refueling_facility_powers_up; Pete Davies, *American Road: The Story of an Epic Transcontinental Journey at the Dawn of the Motor Age* (New York: Henry Holt, 2002); see also Henry Petroski, "On the Road," *American Scientist*, September/October 2006, pp. 396–99.

44. Http://www.fuelcellpartnership.org; Greg Bensinger, "Ford Says Hydrogen Vehicles Not Marketable Until 2030," *Bloomberg.com*, Oct. 3, 2008, http://www.bloomberg.com/apps/news?pid=newsarchive&sid=aE0VQ6_B6C14.

45. Bensinger, "Ford Says"; Jad Mouawad, "Pumping Hydrogen," *New York Times*, Sept. 24, 2008, "Business of Green" special section, pp. 1, 8.

46. "Looking at Hydrogen to Replace Gasoline in Our Cars," *Scientific American*, July 3 2008; see http://www.emagazine.com/earthtalk/archives.php; Reuters, "Study: Fuel Cell Cars Still 15 Years Away at Best," *Los Angeles Times*, July 17, 2008, http://www.latimes.com/news/science/la-sci-fuel18-2008jul18,0,4695785; Bensinger, "Ford Says."

47. Kate Galbraith, "Pickens Plan Stirs Debate, and Qualms," *New York Times*, Aug. 4, 2008, pp. C1, C7; "T. Boone Pickens Rides the Wind," editorial, *New York Times*, July 22, 2008, p. A18.

48. Chesapeake Energy Corporation, "Let's Help Rescue America," display ad, *New York Times*, Sept. 2, 2008, pp. A10–A11; see also Chesapeake Energy Corporation, "Let's Rescue America's Drivers," display ad, *New York Times*, Sept. 9, 2008, pp. A22–A23; http://www.pickensplan.com/theplan; Brown, "Emerging Alternate Energy Consensus," p. 18.

49. Galbraith, "Pickens Plan Stirs Debate," p. C7; Patrick Barta, "Thais Lead Drive to Natural-Gas Cars," *Wall Street Journal*, Oct. 21, 2008, http://online.wsj.com/article/SB122454641299552177.html.html; Greg Migliore, "Toyota Shows Natural-Gas Camry in L.A.," *AutoWeek*, Nov. 19, 2008, http://www.autoweek.com/apps/pbcs.dll/article/20081119/FREE/811199991.

50. Bill Vlasic, "The Huge Hybrid: A New Twist on S.U.V.'s Finds Few Takers," *New York Times*, May 31, 2008, pp. A1, A10; Martin Fackler, "Latest Honda Runs on Hydrogen, Not Petroleum," *New York Times*, June 17, 2008, pp. C1, C5.

51. Richard S. Chang, "The Illusion of Miles Per Gallon," *New York Times*, June 20, 2008, Automobiles Section; "M.P.G. Can Mislead When Searching for Fuel Efficiency," *New York Times*, June 24, 2008, p. F4; see also Ron Lieber and Tara Siegel Bernard, "Ditch the Gas Guzzler? Well, Maybe Not Yet," *New York Times*, Aug. 2, 2008, pp. C1, C4;

Richard P. Larrick and Jack B. Soll, "The MPG Illusion," *Science*, June 20, 2008, pp. 1593–94.

52. Chris Woodyard, "Ford's New Gauges Sprout Leaves," *USA Today*, Oct. 29, 2008, p. 5B.

53. Anand Giridharadas, "How to Build a $2,500 Car," *New York Times*, Jan. 8, 2008, pp. C1, C5; Anand Giridharadas, "Protests Halt India's Plant for Cheapest Car," *New York Times*, Sept. 3, 2008, p. A9; Anand Giridharadas, "Indian Carmaker Says Local Politics Stopped Factory," *New York Times*, Sept. 4, 2008, p. A8; Reuters, "India: Automaker Abandons Site," *New York Times*, Oct. 4, 2008, p. A6; Joe Leahy, "Steel Amid Adversity: Tata After Mumbai," *Financial Times*, Dec. 7, 2008, http://www.fco.cat/files/imatges/Butlleti161/FT.pdf.

CHAPTER 10. COMPLEX SYSTEMS

1. Gore quoted in Richard Morgan, "Beyond Carbon: Scientists Worry About Nitrogen's Effects," *New York Times*, Sept. 2, 2008, p. F3; Al Gore, *An Inconvenient Truth: The Planetary Emergency of Global Warming and What We Can Do About It* (New York: Rodale, 2006); Mark Clayton, "How White Roofs Shine Bright Green," *Christian Science Monitor*, Oct. 3, 2008, http://www.csmonitor.com/centennial/events/monitors-stories/2008/09/how-white-roofs-shine-bright-green.

2. Richard Wolfson, "Fuller's Legacy," letter to the editor, *New Yorker*, July 7 & 14, 2008, p. 7.

3. See, e.g., Robert W. Righter, *The Battle over Hetch Hetchy: America's Most Controversial Dam and the Birth of Modern Environmentalism* (New York: Oxford University Press, 2005), pp. xv–xvi.

4. Ibid., p. xvi; Hetch Hetchy time line, http://www.sierraclub.org/ca/hetchhetchy/timeline.asp.

5. "Federal Officials Order Removal of Maine's Edwards Dam," *U.S. Water News Online*, Dec. 1997, http://www.uswaternews.com/archives/arcsupply/7fedoff12.html; Dave Hogan, "Dam's Removal Reshapes Debate," *Oregonian*, July 2, 1999; http://www.bluefish.org/reshapes.htm.

6. Hogan, "Dam's Removal"; Cap'n Jack Duggins, "Re: Dam Removal," *InsideLine.Net Forum*, July 29, 2005, http://www.insideline.net/cgi-bin/ez-forum.pl?noframes;read=31562.

7. Hogan, "Dam's Removal"; Matthew Preusch, "Blasting of Elk Creek Dam Will Begin," *Oregonian*, July 15, 2008, http://www.oregonlive.com.

8. Felicity Barringer, "Pact Would Open River, Removing Four Dams," *New York Times*, Nov. 14, 2008, p. A18.

9. National Academies, *Understanding and Responding to Climate Change:*

Highlights of National Academies Reports, 2008 edition (Washington, D.C.: National Academies, 2008), p. 2; see also John Tierney, "Politics in the Guise of Pure Science," *New York Times*, Feb. 24, 2009, p. D1.

10. National Academies, *Understanding and Responding*.

11. Ibid., pp. 2, 16.

12. David Biello, "Environmental Catch-22?: Mending Ozone Hole May Worsen Climate Change," *SciAm.com*, June 13, 2008, http://www .tnty.com/2008/06/14/environmental-catch-22-mending-ozone-hole-may-worsen-climate-change.

13. Kenneth Chang, "Globe Grows Darker as Sunshine Diminishes 10% to 37%," *New York Times*, May 13, 2004, p. A22; Kenneth Chang, "Earth Has Become Brighter, But No One Is Certain Why," *New York Times*, May 6, 2005, p. A24; Nova, "Dimming the Sun," http://www .pbs.org/wgbh/nova/sun.

14. David Appell, "The Sun Will Eventually Engulf Earth—Maybe," *Scientific American*, Sept. 2008, http://www.sciam.com/article.cfm?id= the-sun-will-eventually-engulf-earth-maybe.

15. Andrew Revkin quoted in Chris Mooney, *Storm World: Hurricanes, Politics, and the Battle over Global Warming* (Orlando, Fla.: Harcourt, 2007), p. 241. On the prostate-specific antigen (PSA) test, see, e.g., Shannon Brownlee, *Overtreated: Why Too Much Medicine Is Making Us Sicker and Poorer* (New York: Bloomsbury, 2007), pp. 200–202.

16. Mark Kinver, " 'Fewer Hurricanes' as World Warms," *BBC News*, May 18, 2008, http://www.bbc.co.uk/go/pr/fr/-/2/hi/science/nature/ 7404846.stm.

17. Mooney, *Storm World*, pp. 257, 262; James Kanter and Andrew C. Revkin, "World Scientists Near Consensus on Warming," *New York Times*, Jan. 30, 2007, p. A13.

18. Andrew C. Revkin, "Can Global Warming Be Studied Too Much?" *New York Times*, Dec. 3, 2002, p. F1; Scott LaFee, "Leaves of Gas," *SignOnSanDiego.com*, Aug. 28, 2008, http://www.signonsandiego .com/uniontrib/20080828/news_1c28trees.html; Tällberg Foundation, "<350," display ad, *New York Times*, June 23, 2008, p. A7.

19. Beth Daley, "US Castoffs Resuming Dirty Career," *Boston Globe*, Aug. 19, 2007, pp. A1, A22.

20. National Academies, *Understanding and Responding*, p. 20.

21. Alan Zarembo, "It's a Tidy Answer to Global Warming: Giant Vacuums Would Suck up Carbon Dioxide. It Works, but the Cost Is Nearly Out of This World," *Los Angeles Times*, April 29, 2008, p. A15.

22. LaFee, "Leaves of Gas"; Molly Bentley, "Synthetic Trees Could Purify Air," *BBC News*, Feb. 21, 2003, http://news.bbc.co.uk/2/hi/science/

nature/2784227.stm; Michael d'Estries, "Future Tech: Giant Carbon Sucking Trees Might Save the World," *Groovy Green*, June 15, 2007, http://www.groovygreen.com/groove/?p=1500.

23. Zarembo, "It's a Tidy Answer"; Robert Kunzig, "Geoengineering: How to Cool Earth—at a Price," *Scientific American*, Oct. 20, 2008, http://www.sciam.com/article.cfm?id=geoengineering-how-to-cool-earth.

24. Http://www.planktos.com; Cornelia Dean, "Handle with Care," *New York Times*, Aug. 12, 2008, p. F1; Patrick Barry, "Carbon Caveat," *Science News*, Aug. 20, 2008, http://www.sciencenews.org/view/generic/id/35565/title/Carbon_caveat.

25. Kunzig, "Geoengineering"; Clayton, "How White Roofs."

26. Dean, "Handle with Care," pp. F1, F4.

27. Harry Hutchinson, "Taking Engineering's Pulse," *Mechanical Engineering*, June 2008, p. 37.

CHAPTER 11. TWO CULTURES

1. C. P. Snow, *The Two Cultures: And a Second Look* (New York: Mentor, 1964), pp. 10–11.

2. Ibid., pp. 19–20.

3. Ibid., pp. 18, 21.

4. Bruno Latour, *Aramis, or the Love of Technology*, trans. Catherine Porter (Cambridge, Mass.: Harvard University Press, 1996), pp. viii, 23, 24, 210.

5. Ibid., p. 113.

6. Extensive excerpts from the lecture were published in the May 1959 and subsequent issues of *Encounter*; F. R. Leavis, *Two Cultures?: The Significance of C. P. Snow* (New York: Pantheon, 1963), pp. 28, 30; Snow, *The Two Cultures*, p. 7. According to Snow's preface, the second edition of the lecture differs from the first only in "the correction of two small inaccuracies," which he does not identify.

7. Snow, *The Two Cultures*, p. 59.

8. Ibid., pp. 60–65, 74.

9. Ibid., pp. 35–36.

10. Ibid., pp. 64–65.

11. Ibid., p. 27.

12. On the Boeing 777 design process, see Henry Petroski, *Invention by Design: How Engineers Get from Thought to Thing* (Cambridge, Mass.: Harvard University Press, 1996), Chapter 7.

13. Henry Petroski, *To Engineer Is Human: The Role of Failure in Successful Design* (New York: St. Martin's, 1985); paperback edition published in 1992 by Vintage.

CHAPTER 12. UNCERTAIN SCIENCE AND ENGINEERING

1. Natalie Angier, *The Canon: A Whirligig Tour of the Beautiful Basics of Science* (Boston: Houghton Mifflin, 2007), pp. 9, 15.
2. Frank L. Stahl, Daniel E. Mohn, and Mary C. Currie, *The Golden Gate Bridge: Report of the Chief Engineer*, Volume II (San Francisco: Golden Gate Bridge, Highway and Transportation District, 2007), pp. 145–58.
3. "Major Seismic Event in California Predicted," *Civil Engineering*, June 2008, p. 30.
4. Laurie A. Shuster, "ASCE Panel Finds Risk Level in New Orleans 'Unacceptable,' " *Civil Engineering*, June 2008, pp. 15–16.
5. Matthew L. Wald, "Panel Backs Letting Airlines Confess Errors Unpunished," *New York Times*, Sept. 11, 2008, p. A18; "FAA Air Safety Reforms Ordered," *Regional Aviation News*, Sept. 15, 2008.
6. "Corps Said Area Has Never Been Safer," *WDSU.com*, June 5, 2008.
7. B. John Garrick, *Quantifying and Controlling Catastrophic Risks* (San Diego: Academic Press, 2008), pp. 1, 2.
8. Ibid., pp. 3, 5.
9. See, e.g., Mark D. Abkowitz, *Operational Risk Management: A Case Study Approach to Effective Planning and Response* (Hoboken, N.J.: Wiley, 2008), pp. ix–x, 258–68; American Society of Civil Engineers, *The Vision for Civil Engineering in 2025* (Reston, Va.: ASCE, 2007), p. 9.
10. See Diane Vaughan, *The Challenger Launch Decision: Risky Technology, Culture, and Deviance at NASA* (Chicago: University of Chicago Press, 1996), p. 274.
11. Garrick, *Quantifying and Controlling*, p. xxi.
12. Ibid., pp. 3–4.
13. David Appell, "The Sun Will Eventually Engulf Earth—Maybe," *Scientific American*, Sept. 2008, http://www.sciam.com/article.cfm?id=the-sun-will-eventually-engulf-earth-maybe.
14. Garrick, *Quantifying and Controlling*, pp. 2, 91, 95; for tables and supporting documentation associated with the calculations, see Chapter 4 and Appendix C.
15. National Academies, *Understanding and Responding to Climate Change: Highlights of National Academies Reports*, 2008 ed. (Washington, D.C.: National Academies Press, 2008), p. 23.

CHAPTER 13. GREAT ACHIEVEMENTS AND GRAND CHALLENGES

1. George Constable and Bob Somerville, *A Century of Innovation: Twenty Engineering Achievements That Transformed Our Lives* (Washington, D.C.: Joseph Henry Press, 2003).

2. Irwin Goodwin, "Engineers Proclaim Top Achievements of 20th Century, but Neglect Attributing Feats to Roots in Physics," *Physics Today*, May 2000, pp. 48–49.

3. "Northeast Blackout of 1965," http://en.wikipedia.org/wiki/Northeast_Blackout_of_1965; "Northeast Blackout of 2003," http://en.wikipedia.org/wiki/2003_North_America_blackout; Blackout History Project, "Great Northeast Blackout," http://www.blackout.gmu.edu/events/tl1965.html; J. R. Minkel, "The 2003 Northeast Blackout—Five Years Later," *ScientificAmerican.com*, Aug. 13, 2008, http://www.scientificamerican.com/article.cfm?id=2003-blackout-five-years-later.

4. "California Electricity Crisis," http://en.wikipedia.org/wiki/California_electricity_crisis; Congressional Budget Office, "Causes and Lessons of the California Electricity Crisis," Sept. 2001, http://www.cbo.gov/ftpdocs/30xx/doc3062/CaliforniaEnergy.pdf.

5. On the development of improved roads, see I. B. Holley Jr., *The Highway Revolution, 1895–1925: How the United States Got Out of the Mud* (Durham: Carolina Academic Press, 2008).

6. On the impossibility of achieving perfection, see Henry Petroski, *Small Things Considered: Why There Is No Perfect Design* (New York: Alfred A. Knopf, 2003); "The Betamax vs. VHS Format War," *Media College.com*, http://www.mediacollege.com/video/format/compare/betamax-vhs.html.

7. Stuart Schwartzapfel, "Technology Makes Sound More Like Fully Internal Combustion Vehicles," *Wired*, Autopia blog, Aug. 6, 2008.

8. Mary Roach, "Questions for Ray Bradbury; Martian Tourist," *New York Times Magazine*, Nov. 5, 2000, p. 21; James Hibberd, "Ray Bradbury Is on Fire!," *Salon.com*, Aug. 29, 2001, http://www.salon.com/people/feature/2001/08/29/bradbury.

9. Jay Royston, "Denver International Airport: A Case Study," paper prepared for my course Introduction to Structural Engineering, offered during the Spring 2008 semester at Duke University; Paul Stephen Dempsey, Andrew R. Goetz, and Joseph S. Szyliowicz, *Denver International Airport: Lessons Learned* (New York: McGraw-Hill, 1997), p. xi.

10. Glenn Rifkin, "What Really Happened at Denver's Airport," *Forbes*, Aug. 29, 1994, p. 110; Dempsey et al., *Denver International Airport*, pp. xi, 303.

11. Rifkin, "What Really Happened."

12. Neil MacFarquhar, "Trying to Curb Global Heat, U.N. to Turn Up Its Own," *New York Times*, July 31, 2008, p. A6.

13. For a history of air-conditioning, see Gail Cooper, *Air-conditioning America: Engineers and the Controlled Environment, 1900–1960* (Baltimore: Johns Hopkins University Press, 1998).

14. Andrew Jacobs, "Worrying About Traffic, Not Who Wins the Gold," *New York Times*, Aug. 6, 2008, p. A8.

15. National Academies, "Leading Engineers and Scientists Identify Advances That Could Improve Quality of Life Around the World," press release, Feb. 15, 2008; Alan S. Brown, "Academy Unveils Grand Engineering Challenges," *Mechanical Engineering*, April 2008, p. 8; National Academy of Engineering, *Grand Challenges for Engineering* (Washington, D.C.: National Academy of Sciences, 2008).

16. John M. Broder and Matthew L. Wald, "Big Science Role Is Seen in Global Warming Cure," *New York Times*, Feb. 11, 2009, p. A24.

17. Steve Lohr, "Wall Street's Extreme Sport," *New York Times*, Nov. 5, 2008, pp. B1, B5.

18. "After the Crash: How Software Models Doomed the Markets," editorial, *Scientific American*, Nov. 21, 2008, www.sciam.com/article.cfm?id= after-the-crash.

19. Elizabeth Kolbert, "The Island in the Wind," *New Yorker*, July 7 & 14, 2008, pp. 74–77.

CHAPTER 14. PRIZING ENGINEERING

1. Michael Cooper, "McCain Proposes a $300 Million Prize for a Next-Generation Car Battery," *New York Times*, June 24, 2008, p. A20.

2. Ibid.

3. William M. Bulkeley, "The Search for a Better Battery Seems Everlasting," *Wall Street Journal*, Oct. 28, 2008, p. A11.

4. Tom Abate, "Latest in Battle Gear Is a Prize: Batteries," *San Francisco Chronicle*, June 22, 2008, p. C-1; see also William Grimes, "Heard the One About the Farmer's Ethanol?" *New York Times*, March 7, 2008, p. B34.

5. Bulkeley, "The Search for a Better Battery."

6. Duncan Graham-Rowe, "Fifty Years of DARPA: A Surprising History," *New Scientist.com*, May 15, 2008, http://www.newscientist.com/article/ dn13908-fifty-years-of-darpa-a-surprising-history.html.

7. Http://www.nmm.ac.uk/server/show/conWebDoc.355; see also Dava Sobel, *Longitude: The True Story of a Lone Genius Who Solved the Greatest Scientific Problem of His Time* (New York: Penguin, 1996); monetary equivalents calculated at http://eh.net/hmit.

8. Http://www.charleslindbergh.com/plane/orteig.asp; http://en.wikipedia .org/wiki/Spirit_of_St._Louis.

9. On the establishment of the Nobel Prizes, see Elisabeth Crawford, *The*

Beginnings of the Nobel Institution: The Science Prizes, 1901–1915 (Cambridge and Paris: Cambridge University Press/Éditions de la Maison des Sciences de l'Homme, 1984).

10. See, e.g., Henry Petroski, *Remaking the World: Adventures in Engineering* (New York: Alfred A. Knopf, 1997), pp. 12–24.

11. "Doc Draper & His Lab," http://www.draper.com/profile/docslab .html; Henry Petroski, "The Draper Prize," *American Scientist*, March/ April 1994, pp. 114–17.

12. Http://web.mit.edu/invent/a-main.html.

13. "Incentivising Invention," *Economist*, March 1, 2007, http://www .economist.com/world/international/displaystory.cfm?story_id=877941 9; Jane Byrne, "Minds on the Prize," *PE: The Magazine for Professional Engineers*, July 2006, p. 27; John Schwartz, "Up to $25 Million in Prizes Is Offered for an Unmanned Landing on the Moon," *New York Times*, Sept. 14, 2007, p. A14; Tariq Malik, "Space Tourism Firms Set for Big Leaps," *Space.com*, June 20, 2008, http://www.space.com/news/080620-virgingalactic-spaceadventures.html.

14. Schwartz, "Up to $25 Million"; Brad Stone, "A Google Competition, with a Robotic Moon Landing as a Goal," *New York Times*, Feb. 22, 2008, p. C3.

15. Mark Clayton, "Next X Prize: Build a Practical, Hyperefficient Car," *Christian Science Monitor*, April 4, 2007, p. 1.

16. Http://www.progressiveautoxprize.org.

17. Jesse McKinley, "Race Starts with Little Fuel, and Goes Uphill from There," *New York Times*, Oct. 13, 2008, p. A12; Steve Friess, "With Little Fuel, Eco-Racers Arrive in Las Vegas," *New York Times*, Oct. 14, 2008, http://nytimes.com/2008/10/15/us/15eco.html; http:// www.theshipyard.org.

18. Buckminster Fuller Institute, "The Buckminster Fuller Challenge," http://challenge.bfi.org; Bob Kalish, "Bath Man in Pursuit of Futuristic Eco-homes," (Brunswick, Maine) *Times-Record*, Aug. 4, 2008, pp. A1, A12.

19. National Academies, "National Academy of Engineering Announces Winners of $1 Million Challenge to Provide Safe Drinking Water," press release, Feb. 1, 2007; Abul Hussam, Sad Ahamed, and Abul K. M. Munir, "Arsenic Filters for Groundwater in Bangladesh: Toward a Sustainable Solution," *The* (National Academy of Engineering) *Bridge*, Fall 2008, pp. 14–23.

20. Ian Austen, "Solve a Mining Puzzle and Win $10 Million," *New York Times*, Sept. 20, 2007, p. C10.

21. Ibid.

22. Paul Rincon, "US Team Wins Asteroid Competition," *BBC News*,

Feb. 26, 2008; John Johnson, "U.S. Not Prepared for Possible Asteroid Strike, Group Says," *Los Angeles Times*, July 5, 2008, http://www .latimes.com/news/science/la-sci-asteroids-2008jul05,0,478407.story.

23. Rincon, "US Team Wins."
24. Ibid.
25. Elizabeth Kolbert, "The Island in the Wind," *New Yorker*, July 7 & 14, 2008, pp. 68–74.
26. Ibid.

Illustration Credits

13 Trees flattened during the Tunguska Incident. *From the Leonid Kulik Expedition.*

21 Theodore von Kármán commemorative U.S. postage stamps. *From the author's collection.*

49 Isambard Kingdom Brunel. *From the author's collection.*

68 U.S. patent issued to Albert Einstein and Leó Szilárd. *From U.S. Patent No. 1,781,541.*

80 The New York Times Building. *Photo © David Sundberg/Esto; courtesy of Geto & de Milly, Inc.*

96 Photograph of a purported meeting between Albert Einstein and Charles Steinmetz. *Courtesy of Public Affairs Department, General Electric Company.*

98 An assemblage of ampersands from a variety of typefaces. *Composed in Microsoft Office Word typefaces.*

108 Vannevar Bush on the cover of *Time* magazine for April 3, 1944. *From the author's collection.*

120 Research and development expenditures in the United States as a percentage of gross domestic product. *Based on data from the Congressional Budget Office and the National Science Foundation.*

128 Wind turbines in a field. *Photo courtesy of the Pickens Plan.*

187 The Golden Gate Bridge on its fiftieth anniversary. *Photo by John O'Hara; from the author's collection.*

208 A plan for runways at the Denver International Airport. *From* Denver International Airport: Lessons Learned; *used with permission of Paul Stephen Dempsey.*

Index

Italicized page numbers refer to illustrations and their captions.

accidents, aircraft, 8, 191–2, 205
acid rain, 135
Acupuncture Research College of Los
    Angeles, 60
AEG (German electric company), 67,
    68
aerodynamics, 51
air-conditioning, 209–10
airline industry, 191–2, 205
airplane, the, 51, 204–6
airplanes, 53, 54
    see also specific aircraft
air pollution, 76, 212
airports, 206
Alberts, Bruce, 46
Alcoa (aluminum company), 94
    research labs, 93, 94, 99
algae, for biofuel, 146
Allen, Barry, 45, 53, 119
Aluminum Company of America, *see*
    Alcoa
American Rivers (advocacy group), 160
American Society of Civil Engineers,
    194
ammonia (refrigerant), 67
ampersand, 97, 98, *98*, 99, 100
Anderson, Philip W., 103
Angier, Natalie, 36, 184
Ansari X PRIZE, 223
Anschütz-Kaempfe, Hermann, 65
Antheil, George, 59–60

Apollo program, 23, 111, 194, 224
Apollo 11 (mission), 48
Apollo 13 (mission), 55–6, 161, 194–5
Apophis (asteroid), 227–8
applied science, 45, 178
    and engineering, 45, 145
architects, vs. engineers, 21, 30
Argonne National Laboratory, 25, 31–3
Arkin, Ronald C., 35
ARPANET, 57–8
arsenic, in water, 226
artificial heart, 60
artists, 31, 225
Astaire, Fred, 86
asteroids, 14, 192, 196–7, 227
    on collision course with Earth,
        11–16, 28, 227
        preventive measures against, 11,
            12, 14–15, 22, 227–8
    predicting path of, 14, 15, 16, 227
astronauts, 6, 9, 15, 55–6
AT&T, research labs, 94, 98, 99
Atlantic, The, 90
Atlantis (space shuttle), 195
atomic energy, 124–5
    see also nuclear power
Augusta, Maine, 158, 159
automobiles, 54, 198, 201–4, 211, 224–5
    electric, 84–5, 203–4
    see also specific models
Automotive X PRIZE, 224, 225

bacteria, 5, 6, 7
Baker, William F., 78
Ballet Mécanique (Antheil), 60
Bangladesh, water supply in, 226
Bank of Sweden Prize in Economic
    Sciences, 222
Barrick Gold (mining firm), 227
batteries, 85, 137–8, 142, 218–19
Bayh-Dole Act (1980), 115
Bell Telephone Laboratories, 98, 114,
    119, 121
bicycles, 54–5, 84, 225
Big Bang, 33, 91
Billerica, Mass., 150
Bin Laden's Plan (Malone), 82
biofuels, 6–7, 130, 142–5, 146
biologists, 6, 21, 34, 106
biosphere, Earth's, 156
bismuth, liquid, 69
BlackBerry, 90
black boxes (computers), 32–3, 39, 165
Black Hills, S.D., 171
black holes, 10, 33
blackouts, electric power, 200–1
Bloomberg, Michael, 129
BMW (automaker), 150
Board of Longitude, 220
Boeing 747 accident, 205
Boeing 777 (airliner), 172, 180
Bohr, Niels, 24
bombs, atomic, 9, 13, 105, 107, 112,
    113, 124
Bradbury, Ray, 204
brain, reverse-engineer the, 213
Brazil, 146, 153
bridges, design of, 47–8, 53–4, 61,
    175–6
Brooklyn Bridge, 129, 176
Brown, Gordon Stanley, 114
Brownian motion, 62
Brown University, 72
Brunel, Isambard Kingdom, 49, *49*
Buckminster Fuller Challenge, 225–6
Budapest Technical University, 65
"buildering," 79
Burj Dubai (skyscraper), 78

Bush, Vannevar, 105–9, *108*, 111,
    113–14, 115, 118
butane, 67, 148

Calatrava, Santiago, 78
calculators, desk, 57
California energy policies, 131, 134–5,
    201
California Fuel Cell Partnership, 150
California Institute of Technology, 20,
    46, 60, 100
Caltrans, 188
Campobello Island, 139
Camry (Toyota car model), 152
cap-and-trade systems, 135–6
Cape Wind (project), 128
carbon, 7, 143, 144, 145, 170
    dioxide, 72, 122, 130, 131, 135, 146,
        155, 167, 168, 169, 171
    emissions, 86, 129–30, 131, 135–6,
        167, 169
    filters, 55–6, 169
    footprints, 147, 161
    offsets, 170, 171
    sequestration, 146, 169, 213
Carnegie Institution of Washington,
    101–2, 106
Carr, Nicholas, 90
Cascade Engineering (wind turbines),
    131
Cascadia Subduction Zone, 189
catastrophes, 9, 192, 196
Center for Automotive Research, 137
Center for Global Ethics, 171
Challenger (space shuttle), 156, 195
challenges, grand engineering, 212–13
Charles Stark Draper Prize, 222, 223
chemical engineering, 118
chemistry, 4, 22, 42, 52, 62, 102, 118,
    221, 222
chemists, 21
China, 3–4, 119, 179, 180, 212, 215
chlorofluorocarbons, 53, 71, 144, 163–4
    see also Freon
choice, engineering and, 54

Christian Science Monitor, 143
Chrysler (automaker), 138
Chu, Steven, 213
Civic GX (Honda CNG car), 152
civil engineers, 30, 148, 198
Clarke, Arthur C., 14–15, 48
Clarke, Renaldo, 80–1
climate change, 7, 16, 17, 91, 121, 122,
    161–4, 166–7, 168–72, 183, 193,
    196, 197
    see also global warming
Clough, G. Wayne, 30
Club Watt (Rotterdam), 142
coal, 132, 144, 157, 199, 212
    coal-burning plants, 134, 146, 168
    and steamships, 48–9
coho salmon, 160
Columbia (space shuttle), 156, 195
Columbia University, 60
complexity, 61, 144, 156, 171
complex systems, 91, 156–72
compressed natural gas (CNG),
    151–2
    see also natural gas
Compton, Karl, 106
computer codes, 32–3
computer industry, 60
computer models, 214
computers, 42, 56–7, 211
    see also data centers
computer science, 57, 62
Conant, James B., 106
Concorde (supersonic aircraft),
    205–6
conservation, fuel, 147
Copenhagen Business School, 122
copper, 72, 132
corn, and biofuels, 6, 142–3, 145,
    146
Corning, Inc., research budget, 120
Corning Glass Works, 101–2, 115,
    118
    R&D at, 94, 99, 101–4, 115, 116,
    118–19
creative processes, 24, 47, 53
Cross, Hardy, 145

cyberspace, 213
cyclotron, 65

Daily News (New York tabloid), 82
Daimler (automaker), 150
Dallas–Fort Worth Airport, 207
dams, 127, 140, 157–61, 179, 180
    removal of, 158, 159–61
D&R, *see* development and research
Danish Ministry of Environment and
    Energy, 228–9
Darlin, Damon, 90
DARPA Grand Challenge, 220
data centers, 42–3
Davidson, Michael, 89
Day, Arthur L., 101, 102, 103–4
Defense Advanced Research Projects
    Agency (DARPA), 219
De Forest, Lee, 64
Denver, Colo., 206
Denver International Airport, 206–9,
    208
desert tortoise, 134
design, 3, 23, 37, 45, 83, 92, 181, 203
    computer-aided, 211
    as creative process, 47–8, 175–6
    engineering and, 22, 42, 46, 116
    engineers and, 30, 38–9, 41
    and failure, 181, 203
    as hypothesis, 61, 62
    and science, 45, 47
    and unintended consequences, 84
    see also bridges; rockets
Design News, 137
Detroit River, 141
development and research, 116–19, 123
differential analyzer, 106
Dilbert (comic-strip character), 40–1
dinosaurs, extinction of, 12, 16
discovery, vs. invention, 22
DNA, 46
Douglas Aircraft Company, 98
Dover Road (Durham, N.C.), 74, 75
Draper Prize, *see* Charles Stark Draper
    Prize

drug delivery systems, 30
Duke University, 180–2
DuPont (company), 118, 219
  R&D at, 94, 99
dymaxion, 55

Earth, threats to, 9, 10, 11–16
  see also specific threats
earthquakes, 12, 185–90, *187*, 192, 196,
    228
  and bridges, 186–90
Eastman Kodak, *see* Kodak
East St. Louis, Ill., 226
E. coli contamination, 4
economics, 48, 118, 121–2, 134, 170,
    222
Economist, The, 121, 223
economists, 119, 122, 143, 146, 197,
    214
Edison, Thomas, 85, 93, 95
Edison Electric Light Company, 93
Edison General Electric Company, 93
Edison Illuminating Company, 85
Edwards Dam (Maine), 158–60, 161
Edwards Manufacturing Company,
    158–9
E85 (biofuel), 143
  see also ethanol
Einstein, Albert, 21, 24, 31, 63, *96*, 105
  on invention, 21–2,
  as inventor, 62–3, 65, 66–9, 70
  as patent examiner, 63–4, 65
  and patents, 63–5, 66, *68*
  and refrigerator, 66, 67–9, *68*
  and Steinmetz, 95–7, *96*
  see also Szilárd, Leó, and Einstein
Einstein-Szilárd refrigerator, 66, 67–9,
    68, 70, 71
Eisenhower, Dwight D., 150
electrical engineering, 60, 63
electricity, 85, 126, *128*, 147, 199, 226
  generation of, 127–42, 145, 148, 152,
    157–8
  from nuclear power, 124–5, 127,
    132

transmission of, 126, 129, 130
  see also blackouts; fuel cells; solar;
    wind
electrification, 126, 198–9, 211
Electrolux Servel Corporation, 67
electromagnetic pump, 69
electromagnetic waves, 50, 64, 114
Elk Creek Dam (Ore.), 160
Elsinore fault, 189
e-mail, 88–90
Empire State Building, 83
Encounter (magazine), 177
energy, 130, 138–40
  alternative sources of, 121, 127–54
  density, 85, 137
  from fossil fuels, 130
  from the ocean, 138–40
  renewable, 130
  schemes to save, 71–2
  sustainable use of, 121
  see also hydroelectric power; solar
    power; wind
Energy Independence and Security Act
    (2007), 71
engineer (as word), 36,
Engineer and His Profession, The
    (Kemper), 28
engineering, 118, 152, 155, 174, 194,
    199–200
  as creative endeavor, 175–6
  education, 37, 41–2, 117–18, 179,
    180
  great achievements in, 198, 199, 210,
    211
  international nature of, 179–80
  leading science, 45, 47, 49–58,
    109–11, 199
  method, 51, 62
  not applied science, 45, 109
  and science, 52, 110, 183
  vs. science, 17, 20–1, 22, 24–5,
    45–58, 69–70, 113, 175, 178,
    183, 199
  scientists, 21, 24, 38
  study of, 179
  as word, 174

engineers, 23, 34, 176, 194, 198–9, 210
  and fixing things, 37, 38
  jokes about, 38, 90–1
  and medical devices, 30
  as optimists, 90–1
  personality of, 38, 40–1
  relative status of, 26, 29, 30
  vs. scientists, 17, 19–43, 90, 91, 95–6,
    96, 172, 178, 185
  see also specific kinds of engineers
entropy, 90, 91
environmental costs, 211
environmental engineering, 146
environmental engineers, 30
environmental hazards, 7, 72
environmental impact, 133–4, 141,
  143–4, 212
  studies, 146, 160
environmentalists, 80, 132, 136, 144,
  149, 158, 160
environmental scientists, 170
Equinox (Chevy fuel-cell vehicle), 151
Escape from Berkeley (road race), 225
ethanol, 142–3, 144, 146
ethics, 134, 171
European Organization for Nuclear
  Research (CERN), 9, 10
European Union, 7, 71

failure, 45, 61, 73, 181, 194–5, 203, 221
  success and, 24, 28, 71, 97, 105, 181,
    194–5
failures, 27, 28, 99, 110, 181, 201
FCX Clarity (Honda fuel-cell car), 151
Fermi, Enrico, 24
Feynman, Richard, 57, 138
financial crisis, 130, 214
financial engineering, 214
Finnegans Wake (Joyce), 34
fish, and dams, 159, 160, 161
Florida, 5, 45
  and ocean energy, 140
food, tainted, 3–5
Ford, Henry, 85, 149
Ford (automaker), 151, 152, 153

Foresight (spacecraft), 228
fossil-fuel power plants, 124, 130, 147,
  199, 211
fossil fuels, 7, 130, 142, 145, 146, 148,
  150, 151, 157, 161, 213, 229
Franklin, Benjamin, 218
Freedom Tower (New York), 83, 129
Free Flow Power (firm), 141
Freon, 69, 71, 163–4
fuel cells, 122, 148, 149, 219
fuel efficiency, 153–4
Fuller, R. Buckminster, 55, 156
fusion, energy from, 212
Fusion (Ford hybrid sedan), 153
Fusion-io (start-up company), 62

Galileo, 109–10, 175
gallium, 132
Garrick, B. John, 193, 195–6
gasoline, 85, 142, 143,
  mileage, 147, 153,
  prices, 144, 146–7, 153
  stations, 86, 149, 150–1, 202
  vehicles powered by, 85, 153
Gates, Bill, 61
Gell-Mann, Murray, 34
General Electric Company, 67, 72, 93,
  94–5, 98, 101, 106, 121
  R&D at, 94–5, 98, 99, 101, 121
General Motors (automaker), 137–8,
  150, 151, 152
geoengineering, 170–1
George Mason University, 171
Georgia Institute of Technology, 30,
  142
geothermal energy, 136
Geothermal Energy Association, 137
Germany, 67–8, 102, 103, 133
glass, 103, 118–19
  borosilicate, 103
  technology, 102
"global dimming," 164–5
global positioning system (GPS), 204
Global Summit on the Future of
  Mechanical Engineering, 171

global warming, 80, 143–4, 156, 161,
        162–3, 164, 165
    see also climate change
Goddard, Robert, 23–4
Goddard Institute for Space Studies
        (NASA), 167
Goldberger, Paul, 79
Goldcorp (mining firm), 226
Golden Gate Bridge, 186–8, *187*, 189
Google (Internet firm), 89, 90, 121
Google Lunar X PRIZE, 223–5
Gore, Al, 144, 156
Grainger Challenge Gold Award, 226
Grainger Challenge Prize for
        Sustainability, 226
Grand Challenges (NAE), 212–13
gravity, 157
Great Achievements (NAE), 198, 211
Great Eastern (steamship), 49, *49*
Great Western (steamship), 49
greenhouse gases, 43, 76, 84, 121, 136,
        145, 146, 152, 161, 164, 167,
        168, 211, 213
Guatemala, use of recycled coal plant
        in, 168
Guggenheim Aeronautical Laboratory
        (Caltech), 20
Gulf Stream, 139–40

hadrons, 9
Hammer, Edward E., 72
Hansen, James, 167
Harrison, John, 220
Harvard University, 15, 106
headlines, newspaper, 18–19, 26, 27,
        88–9, 164–5
health informatics, 213
Heimlich, Henry J., 60
Hertz, Heinrich, 50
Hetch Hetchy Valley, 158
highways, 198, 202, 204
    see also interstate highway system
Holbrook, George, 118
Home Depot, The (retail chain), 72
Honda (automaker), 150, 151, 152

Hood, Leroy, 46
Hoover Dam, 127
horses, 84, 86
hospitals, error and risk in, 7–8
Houghton, Alanson, 102, 103
Houghton, Amory, Jr., 101, 102
Houghton, Arthur, 102, 103
How We Invented the Airplane (Wright),
        51
Hubble Space Telescope, 195
Hughes, Thomas P., 22, 63, 64
Human Genome Project, 46
humanities, 172, 174, 179, 180, 182–3
    see also "two cultures, the"
Hurricane Katrina, 190–1
hurricanes, 16, 186, 190, 192–3, 200,
        228
    predictions of, 165, 166, 184–5
Hurt, Robert H., 72
Huxley, Thomas, 33
hybrid vehicles, 147, 149, 152–3, 203–4
    plug-in, 137–8, 218
    see also specific models
hydroelectric power, 127, 130, 136,
        140, 157–9, 199, 201
    see also dams; *specific dams*
hydrogen, 67, 148, 149, 150–1
    filling stations, 149, 150–1
    vehicles powered by, 147–8, 150–1
    see also fuel cells,
hydrokinetics, 140–1
hypothesis, 21, 25, 36–7, 61, 62, 163

IBM (corporation), 89, 119
ICBM (missile), 52
Iceland, renewable energy in, 136–7
Impey, Chris, 35
Inconvenient Truth, An (Gore), 156
India, 119, 154–5, 180, 215, 226
indium, 132
Indonesia, 145
Industrial Revolution, 177–8, 179, 182
    pre–Industrial Revolution, 167
Information Overload Research Group,
        89

infrastructure, urban, 213
Institute of Medicine, 8, 112, 120
integrated circuit, 222
Intel (firm), 89
Intergovernmental Panel on Climate
    Change, 166
internal combustion engine, 84, 85,
    126, 143, 149, 151, 152, 203
International Astronomical Union, 16
International Space Station, 223
Internet, 42, 43, 57, 58, 90
interstate highway system, 150
invention, 3, 22, 58, 137, 145, 220, 223
    and science, 47, 51, 53, 64, 69
    see also specific inventions and inventors
inventions, 22, 58–61, 71, 204, 221
    and patents, 115, 120
    see also specific inventions
inventors, 22, 24, 53, 58–61, 64, 66, 69,
    74, 88, 203, 221, 223
    see also individual inventors
ionosphere, 51
Iowa, and biofuels, 143
ISIS (neutron source), 46

Jansky, Karl, 114
Jarvik heart, 60
Jet Propulsion Laboratory, 13, 15, 18,
    20
Jobs, Steve, 61, 86
Johnson, Natalie, 72
Joint Committee on New Weapons and
    Equipment, 106
Jordan, Lewis, 18, 19
journalists, 27, 40
    science, 19, 26, 36, 47
Joyce, James, 34
Jupiter, 11, 15

Kamen, Dean, 86
Kármán vortex street, 20
Kennebec River (Maine), 158, 159
Kennedy, John F., 20
Kettering, Charles F., 59, 60

Kia (automaker), 150
Kilby, Jack S., 222
Kilgore, Harley, 107, 109
King Kong (giant ape), 83
Klamath River, 160
Kline, Ronald R., 94, 98, 99
"knack, the," 40–1
"Knack, The" (video), 40–1
Kodak, R&D at, 94, 99, 100–1, 103
Kulik, Leonid Alekseyevich, 12
Kurly Tie (binding device), 69

Lamarr, Hedy, 59–60
laptop computer, 56–7, 148
Lardner, Dionysius, 48
Large Hadron Collider, 9–11, 31, 91
Lassman, Thomas C., 94, 98, 99
Latour, Bruno, 175–6
Leavis, F. R., 177
LEDs, 73–4, 93
Lee, Wen Ho, 18–19, 24, 26
Lemelson-MIT Prize, 223
levees, 190–1, 192
Lexus LX 570 (SUV), 154
Life Science Library, 39
Light, Andrew, 171
lightbulbs, 71–2, 93, 101
    compact fluorescent, 71–3, 74
Lindbergh, Charles, 221
liquid metal, 69
literary intellectuals, 173, 179
Lomborg, Bjørn, 122
longitude, prize for determining, 220
Los Alamos National Laboratory, 18

Mach, Ernst, 21–2
mad scientist, 35–6
Mahoney, Jerry (dummy), 60
Maine, forest products industry in,
    146
Malaysia, 145, 180
Malone, David, 82
Manhattan Project, 57, 69, 216
    scientists and, 25, 105, 111, 124

Mansfield Amendment (1973), 113
Marburg, Germany, 133
Marconi, Guglielmo, 50–1, 95
Mars, 11, 27
Marshall Field's (department store), 55
Maschinchen (Einstein), 62–3
mathematical physicists, 56
mathematicians, 56, 222
mathematics, 38, 39, 41, 42, 57, 62,
    114, 118, 203
  problem in, 23, 24, 25, 56, 165
Maxwell's Demon, 65
McCain, John, 218, 219
McHenry, David C., Jr., 69
mechanical engineering, 19, 23, 171,
    210
mechanical engineers, 32, 43, 198
medical devices, 30, 41
medical doctors, 30, 217
medical errors, 8
melamine, 4
Menlo Park, N.J., 93
mercury, 72
Merrymeeting Bay (Maine), 159
meteorite, 12
meteor, 11
  see also asteroids
methane, 144, 148
methanol, 148
Michigan, Upper Peninsula of, 129
microbiology, 111
microorganisms, 29
Microsoft (corporation), 89, 120–1
Milan (Mercury hybrid sedan), 153
mileage, fuel, 143, 147
milk, tainted, 3–4
Millikan, Robert, 100
mining prizes, 226–7
Minnesota, and biofuels, 143
mirror, giant, 170
Mississippi River, 141
MIT, 106, 114
  see also Radiation Laboratory
    (MIT)
Model K (Ford car model), 85
Model S (Tesla car model), 138

models, 55, 60, 184
  computer, 32–3, 164, 165, 166, 184,
    214
  mathematical, 56, 214
Model T (Ford car model), 85
Mojave ground squirrel, 134
"Monolith, the," 46
Moon, 11, 52–3, 223–4
  missions to, 48, 52–3, 55–6
    U.S., 9, 52–3, 194–5, 216
      see also Apollo program
  rockets to, 19, 22–3
mud flap, fuel-saving, 147
Muir, John, 158
Munters, Carl Georg, 67

Nano (small Indian car), 154
nanogenerator, 142
nanoparticles, 7, 72, 132
nanoselenium, 72, 132
nanotechnology, 7, 172
Nantucket Island (Mass.), 128, 129
NASA, 20, 27, 111, 121
  and asteroids, 11–12, 13, 14
  see also asteroids
National Academies, 120, 162, 168, 197
National Academy of Engineering, 29,
    62, 112, 120, 171, 174, 198, 211,
    212, 216, 222
  and Grand Challenges, 212–15, 217
  and Great Achievements, 198–211
National Academy of Sciences, 62, 107,
    112, 120, 174
National Advisory Committee for
    Aeronautics (NACA), 106
National Bureau of Standards, 100
National Clean Energy Summit, 129
National Defense Research
    Committee, 107
National Institutes of Health, 109
National Inventors Council, 59, 60
National Oceanic and Atmospheric
    Administration, 166
National Renewable Energy
    Laboratory, 136

National Research Council, 107, 116, 120

National Science and Engineering Foundation (proposed), 112

National Science Board (NSB), 112

National Science Foundation, 107, 109, 111, 112, 117

natural gas, 67, 69, 130, 151–2, 153, 157

natural scientists, as pessimists, 90–1

Near-Earth Object Program (NASA), 13, 14

Netherlands, 127, 145

neutrons, 46

Nevada, and solar power, 135

New Brunswick, N.J., 95

Newcomen, Thomas, 50

New Jersey, 93, 148
  and energy, 128–9, 135

New Jersey Board of Public Utilities, 149

New Orleans, La., 190–1, 192

newspaper reporting, 19, 26–8, 60
  see also headlines, newspaper

Newton's First Law, 185

New York City, 78, 81–3, 228–9, 140
  use of Segways in, 87, 88
  see also specific structures

New Yorker, The, 78

New York Times, The, 5–6, 14, 16, 26, 60, 79, 82, 88, 90, 107

New York Times Building, 78–82, *80*, 83
  scaled by climbers, 79–83

Nissan (automaker), 150

nitrogen, 4, 144, 213

Nobel, Alfred, 221

Nobel Foundation, 221, 222

Nobel Prizes, 52, 119, 213, 221–2
  winners of, 65, 100, 120

"no-regrets" strategies, 168–9

Noyce, Robert N., 222

nuclear physicists, 34

nuclear physics, 24–5

nuclear power, 124–6, 127, 130, 132
  plants, 124–6, 128–9, 134, 139

nuclear reactors, 25, 32, 69, 124

nuclear technologies, 198, 215

nuclear terror, 213, 215

nuclear war, 192, 194

nuclear weapons, 11, 18, 19, 125

Nutting, Perley G., 100–1

Obama, Barack, 147

Oberth, Herman, 24

ocean, 170–1
  energy from, 139–40

October Sky (movie), 24

Office of Naval Research, 109

Office of Scientific Research and Development (OSRD), 107, 109

O'Hare International Airport (Chicago), 207

Ohio River, 141

oil, 6, 126, 130, 142, 151, 218, 224
  crises, 72, 125, 126, 127

"Old Man River's City" (Fuller), 226

Oppenheimer, J. Robert, 124

OptiSolar (solar plant builder), 134–5

Orteig, Raymond, 221

Orteig Prize, 221

O'Shaughnessy Dam, 158

ozone layer, 53, 70, 161, 163–4

Pacific Gas & Electric (utility), 134, 158

PacifiCorp (utility), 160–1

packet switching, 57

Pais, Abraham, 66

palm oil, 145, 146

Panama Canal, 124

Passamaquoddy Bay, 139

Pasteur, Louis, 111

patents, 51, 54, 59, 60, 61, *68*, 69
  international, 64–5, 66, 67
  research and, 107, 115, 120
  see also Einstein, and patents; *specific inventions and inventors*

Pathfinder missions (NASA), 27
Petit, Philippe, 83
Philips Lighting (manufacturer), 73
photoelectric effect, 65
photovoltaic cells, 132, 133
physical chemistry, center of, 102
physicists, 9, 31, 33, 50, 103, 105, 106,
    169, 214
  see also specific individuals and kinds
physics, 42, 48, 52, 57, 62, 64, 112,
    221
  advanced by engineering, 199
  laws of, 22, 25, 47
  vs. other sciences, 33
  see also specific kinds
Piano, Renzo, 78, 82
Pickens, T. Boone, *128*, 151–2
piezoelectric effect, 142
Planetary Society, 227
Planktos (firm), 170
platinum, used in fuel cells, 151
Poldhu (Cornwall, England), 50
politics, art of, 53
pollutants, global warming, 144
pollution, 7, 30, 44, 74, 76, 84, 88, 145,
    167–8, 212
"pollution markets," 135
power lines, 200–1
Price-Anderson Act (1957), 125
Prius (Toyota hybrid vehicle), 149
prizes, 218, 220–1, 229
  engineering, 222–3
  *see also specific competitions, contests,
    prizes*
probabilities, 162–3, 190, 193
  of earthquakes, 189–90
  of space shuttle failure, 195
probability, 186, 194, 196
  events of low, 11, 16, 191, 193–4,
    196–7
Progressive Automotive X PRIZE,
    224–5
Progressive Insurance Company, 224
Project Hindsight, 113
proof, mathematical, 37, 61
propane, 148

propellers, airplane, 51
prostate cancer, 165–6
protons, 9
Pulitzer Prizes, 222
Pupin, Michael, 64
Pyrex (glass cookware), 103

Q-Cells (solar cell manufacturer), 133
quantitative risk assessment, 163, 193
"quants," 214
quark, naming of, 34
Quebec Bridge, 176

radar, 107
radiation, use on foods, 5–6
Radiation Laboratory (MIT), 107
radio, 50–1, 53, 198
  astronomy, 114
Radio Corporation of America (RCA),
    95
railroads, 179
Ramo, Simon, 52
R&D, *see* research and development
RAND Corporation, 98, 119–20
rapeseed, 146
Rede Lecture (Cambridge), 177
refrigeration, 198
refrigerators, 64, 66–9, 70, 71, 209
  see also Einstein-Szilárd refrigerator
research, 107
  basic, 94, 99, 100, 109, 111, 112–13,
    117, 119, 120–1, 123
  industrial, 102–3, 113, 115, 119,
    120–1, *120*
  medical, 108, 109
research and development, 93–4, 97,
    98, 105, 111–23
  in academic setting, 106, 107, 109,
    117–18
  budgets, 107, 112, 115, 116
  funding of, 107, 111, 116–17,
    119–21, *120*
  linear model of, *108*, 109, 111, 113,
    115, 117, 118

see also development and research; *specific firms and laboratories*
Research Applied to National Needs (NSF program), 112
"Rich and the Poor, The" (Snow), 178
risk, 3–17, 23, 124, 125, 185, 193, 195, 221
  business, 138, 144
  financial, 214
  management, 17, 193, 194
  quantifying, 166, 193, 195–6
  reducing, 72, 81
  spreading, 214
  "unacceptable," 191
  see also quantitative risk assessment
risks, 17, 144, 193
  catastrophic, 193, 195–6
Roadster (Tesla car model), 138
Robert, Alain, 79–80, 81
Rocket Boys (Hickam), 24
rockets, 22–4, 52, 53, 57, 111
rocket science, 22, 52
Rocky Mountains, 207
Roebling, John A., 176
Rogue River (Ore.), 160
rooftops, 131, 171
Roosevelt, Franklin D., 105, 107, 109
  tidal power project, 139
Roosevelt Island (N.Y.), 140
Rutan, Burt, 223
Rutherford, Ernest, 33

Saffo, Paul, 90
Sagan, Carl, 30–1
Sahara Desert, 171
St. Croix River, 139
St. John's, Newfoundland, 50
Salk, Jonas E., 19
salmonella, 5, 6
Samsø, Denmark, 229
San Francisco, 87, 150, 158, 186
San Francisco–Oakland Bay Bridge, 188
San Gorgonio Pass (Calif.), 127
sanitary engineers, 29–30

satellites, artificial, 48, 170, 204
  first Earth, 19, 41, 111, 174
  see also Sputnik; Telstar
Schenectady, N.Y., 94, 121
science, 3, 35, 94, 99
  vs. engineering, 17, 19–21, 25, 45–58, 175, 178
  policy, 107–8
  pure, 31, 94, 145, 178
  and technology, 3, 4, 22, 107, 178
  see also applied science
Science (magazine), 46
Science and Engineering Indicators (NSB), 112
Science Drive (Duke University), 180, 182
science-fiction writers, 48, 52
  see also specific writers
Science Indicators (NSB), 112
Science, Technology, and Human Values (Duke program), 181
Science—the Endless Frontier (Bush), 109, 111, 117
Scientific American, 214
scientific method, 51, 61, 62
scientists, 16, 36, 173
  caricatures of, 34–6
  and R&D policy, 105, 111
  and World War II, 105–6, 111, 124
  see also engineers, vs. scientists; *specific kinds of scientists*
scooters, electric-powered, 86–7
Second Law of Thermodynamics, 65, 173
  see also thermodynamics
"Secret Communication System" (patent), 59
Segway (personal transporter), 86–8
selenium, 72, 132
Sequest (alternative energy company), 146
sequestration, 146, 169, 212
server, computer, 42, 43
Shakespeare, William, 173, 176
Shipyard Labs, 225
Shirley, Donna, 27

Shoemaker-Levy 9 (comet), 11, 12
Sierra Club, 158
Silicon Valley, 43, 89
silver, 7, 227
Singapore Airlines, 205
Skidmore, Owings & Merrill (firm), 78
Skinner, Charles E., 99–101
slide rule, 210–11
Smart Fuel Cell (German firm), 219
Smiff, Knucklehead (dummy), 60
Smithsonian Institution, 30
Snow, Charles Percy, 173, 176–9,
    182–3
    see also "two cultures, the"
social sciences, 107, 172, 183
sodium, liquid, 69
Sojourner (Mars rover), 27
solar cells, 131–2, 133
solar energy, 121, 134–5, 148, 212, 215,
    229
solar panels, 131–3, 135, 136, 138, 148
solar power, 127, 130, 131–5, 140, 170,
    213
sonnets, 176
South Atlantic Ocean, 170
Soviet Union, 19, 41, 111, 174, 219–20
Space Adventures (firm), 223
Spaceship Earth, 156–7, 161
SpaceShipOne (reusable spaceliner),
    223
SpaceShipTwo, 223
space shuttle missions, 156, 195
    see also specific craft
SpaceWorks Engineering (firm), 228
speed bumps, 75, 76–7, 78, 84, 89,
    91–2
speed humps, 74–7, 78
Spirit of St. Louis (airplane), 221
Sputnik (Soviet artificial satellite), 41,
    111, 174, 220
Stanford University, 60
Stanley Steamer (automobile), 201
Stapleton Airport (Denver), 206
    see also Denver International
        Airport
steam-driven automobiles, 84

steam engine, 50, 53, 111, 126, 202
steamships, 48–9
Steinmetz, Charles P., 94–6, *96*
stem cell research, 26–7
Stever, H. Guy, 199
Stony Tunguska River (Siberia), *13*
stream turbines, 139, 140, 141
Strizki, Mike, 148–9
structural engineering, 45, 186
success, *see* failure, success and
Sudan, asteroid impact over, 14
sugar, used making ethanol, 145
Sullivan, Eugene C., 102, 103
sun, destroying Earth, 165, 196
sunflower oil, 146
sunlight, 164, 170
SunPower (solar panel maker), 134
Super Battery Prize, 220
sustainability, 212, 215, 226
Sustainable Dance Club, 142
SUVs (sport utility vehicles), 153
Swift Wind Turbine, 131
Swiss Federal Institute of Technology,
    215
Swiss Federal Polytechnic School, 63
systems, 77, 98, 145, 146, 161, 168–9,
    171
    see also complex systems
Szilárd, Leó, 65, *68*
    and Einstein, 65, 66–9
    see also Einstein-Szilárd refrigerator

Tacoma Narrows Bridge, 20, 176
Tata Motors (Indian automaker), 154
Technische Hochschule, Berlin-
        Charlottenburg, 65
technology:
    importance of, to science, 46, 47,
        111, 114, 178
    history of, 49, 111
telescope, 22, 109–10
Telstar (satellite), 114
Tesla, Nikola, 64
Tesla Motors (electric car company),
    138

thermodynamics, 50, *65*, 66, 69, 114, 229
  see also Second Law of Thermodynamics
Three Gorges Dam, 157
Three Mile Island (accident), 125, 127
Throop Polytechnic Institute, 100
tidal power, 138–9
tide mills, 139
Time (magazine), 107, *108*
Tioga Pass (Sierra Nevada), 225
To Engineer Is Human (Petroski), 181–2
tomatoes, tainted, 5
Torino Impact Hazard Scale, 16, 227
Toyota (automaker), 149, 150, 152
  see also Prius
transatlantic ship crossings, 48–9
trees, synthetic, 169
"trimtab principle" (Fuller), 226
Truman, Harry, 109
tsunamis, 193
Tufts College, 106
Tunguska Incident, 12–13, *13*, 16
turbine blades, 131, 140
  see also stream turbines; wind turbines
Turners Falls, Mass., 168
"two cultures, the," 173–4, 177–83
  and engineering, 174, 178
"Two Cultures, The" (Snow), 177, 179
2,000 Watt Society, 215–16
typhoid fever, 29

U.S. Department of Defense, 109, 113, 219
U.S. Department of Energy, 125, 130
U.S. Energy Research and Development Administration, 125
U.S. Environmental Protection Agency, 125, 143, 192
U.S. Federal Aviation Administration, 8, 191
U.S. Federal Energy Administration, 125
U.S. Federal Energy Regulatory Commission, 159
U.S. Federal Science and Technology Budget, 116
U.S. Food and Drug Administration, 5
U.S. Geological Survey, 101, 189
U.S. Navy, 51
U.S. Nuclear Regulatory Commission, 125, 192
U.S. Securities and Exchange Commission, 214
U.S. Supreme Court, 158
University of Berlin, 65
University of Cambridge, 31, 173, 177
University of Chicago, 100
University of Michigan, 102
University of Utah, 60
uranium, 105

uncertainty, 185–6
Understanding and Responding to Climate Change (National Academies), 162
uniqueness, 56
  lack of, in engineering, 23, 53–4, 56–7
United Nations, 166, 209
U.S. Army, 35, 98, 150, 190–1
U.S. Atomic Energy Commission, 109, 125
U.S. Bureau of Land Management, 134
U.S. Congress, 12, 132

vehicles, motor, 75–7, 84–6, 87, 147, 168, 203–4, 224–5
  all-electric, 84–6, 137, 138, 147
  fuel-cell, 147–8, 149–50, 151
  gasoline-powered, 85, 151, 153–4
  hybrid, 137–8, 147
  natural gas, 151–2
  see also specific models
Verdant Power, 140
Verne, Jules, 48
Vest, Charles M., 171–2
videocassette recorders, 203
Villa Nueva, Guatemala, 168

Virgin Galactic (space tourism firm), 223
virtual reality, 213
VIVACE (hydrokinetic system), 141
volcanoes, 102, 170, 192
Volkswagen (automaker), 150, 154
Volt (Chevy plug-in hybrid), 137–8
von Kármán, Theodore, 20–1, *21*, 26
  on science vs. engineering, 20, 22
von Platen, Carl, 67
von Platen–Munters absorption refrigerator, 67

Wall Street, 214
Wal-Mart (retail chain), 72
Washington (state), 140–1
water, 130, 213
  see also hydroelectric power
Watt, James, 50
waves, ocean, energy from, 139
weapons, 106–7, 109
  see also bombs, atomic
Wearable Power Prize, 219
Weather.com, 184
weeds, 6, 146
Westinghouse Electric Company, 94, 95, 101
  R&D at, 94, 99–101, 104, 112–13
West Orange, N.J., 93
WhiteKnightTwo (mothership), 223
Williams, William Carlos, 24
Winchell, Paul, 60
wind, 126, 128, 129, 130, 131, 157, 207
  and bridges, 175, 176
  energy, 130, 136, 148, 152

farms, 127, *128*, 129, 130, 131, 152
mills, 126, 127, 129
power, 126, 127, 130, 131, 139, 140, 142
turbine business, 121, 127, 130, 140
turbines, 127–8, *128*, 129–31, 139, 228, 229
winds, 48, 161, 164, 190
"windshield wiper effect," 165–6
wireless telegraphy signals, 50
Working Group on California Earthquake Probabilities, 189
World Bank, 143
World Trade Center (New York), 83, 129
  Twin Towers, 82, 83
  see also Freedom Tower
World War I, 85, 103
World War II, 99, 105, 112, 114, 117
  scientists and engineers in, 105–6
World Wide Web, 42, 43, 58, 90, 202
Wozniak, Steve, 62
Wright, Orville, 51
Wright brothers, 51, 55, 58, 204
writing, 175–6
Wulf, William A., 171

X Prize Foundation, 223
X Prizes, 223–5

Yosemite Valley, 158

Zurich Polytechnic, 63